The Fireproof Building

STUDIES IN INDUSTRY AND SOCIETY

Philip B. Scranton, Series Editor

*Published with the assistance of the
Hagley Museum and Library*

Related titles in the series:

David A. Hounshell, *From the American
System to Mass Production, 1800–1932:
The Development of Manufacturing
Technology in the United States*

Lindy Biggs, *The Rational Factory:
Architecture, Technology, and Work
in America's Age of Mass Production*

Mark Aldrich, *Safety First:
Technology, Labor, and Business
in the Building of American
Work Safety, 1870–1939*

THE FIREPROOF BUILDING

Technology and Public Safety
in the Nineteenth-Century
American City

Sara E. Wermiel

The
Johns Hopkins
University Press
Baltimore
& London

© 2000 The Johns Hopkins University Press
All rights reserved. Published 2000
Printed in the United States of America
on acid-free paper

9 8 7 6 5 4 3 2 1

The Johns Hopkins University Press
2715 North Charles Street
Baltimore, Maryland 21218-4363
www.press.jhu.edu

Library of Congress Cataloging-in-Publication
Data will be found at the end of this book.
A catalog record for this book is available
from the British Library.

ISBN 0-8018-6311-2

Contents

Acknowledgments

It is with pleasure that I acknowledge the assistance I have received from many people while doing research for and writing this book. In the endnotes, I acknowledge the individuals who kindly answered questions or sent me material. But I do not even know the names of the building owners, superintendents, managers, curators, and tenants who let me into their basements, offices, apartments, and on to the roofs so that I could see how their buildings were put together; to all of them, my sincere thanks.

Thanks to the librarians at the Massachusetts Institute of Technology, especially those at the Retrospective Collection (RSC), Rotch Library, Institute Archives, and Interlibrary Loan service for all their friendly help. I am particularly obliged to the RSC for letting me keep its books for long periods.

Through a Charles E. Peterson Research Fellowship, awarded by the Athenaeum of Philadelphia, I was able to spend time at the Athenaeum, using its excellent architecture and history collection. Bruce Laverty, the Athenaeum's Curator of Architecture, helped me find material and introduced me to the staff at Girard College. Thanks to the librarians at several membership organizations—the American Institute of Architects, American Insurance Association, National Fire Protection Association, and Insurance Library Association (Boston)—for allowing me to use their private libraries. Sarah Turner, archivist of the Architect of the Capitol's office; the archivists at the National Archives in Washington, D.C.; and the manuscript curators at the Hagley Library were all very helpful. Many thanks to these individuals and the keepers of collections at the other historical societies, museums, and government collection departments in public libraries that I have used and contacted over the years.

A William F. Sullivan Fellowship with the American Textile History Museum supported my research on slow-burning construction. A Society for the History of Technology/Historic American Engineering Record fellowship supported my overall research. At the time that I was exploring various topics for my doctoral dissertation, I received a graduate student fellowship from the Smithsonian Institution, which allowed me to do research at the National Museum of American History. Thanks to these organizations for their support.

In writing my doctoral dissertation, I was fortunate to have a top-flight committee—Robert Fogelson, Merritt Roe Smith, and Peter Temin—that gave me good advice. Three hardy individuals read a draft of this book: Lawrance Hurst, Arnold Pacey, and Francis Ventre. I particularly wish to thank Mr. Hurst and Prof. Pacey for sharing their considerable knowledge about construction history and sources, as well as photographs and drawings, and for introducing me to many interesting buildings in Great Britain. I received helpful comments from individuals who read chapters: Deborah Fitzgerald (chapter 1); Todd Shallat and Martin Reuss (chapter 2); and Mike Williams (chapter 4). In addition, I have benefited from conversations and correspondence with Margot Gayle, Charles Peterson, and Derek Trelstad. Hope Haff, a true friend, listened when I went on about my latest discovery about fireproof buildings. Thanks to everyone.

The Fireproof Building

Introduction

SOLVING THE
PROBLEM OF URBAN
CONFLAGRATION

Sweeping fires are so unusual in American cities today that the once dreaded word *conflagration* sounds quaint to modern ears. This relative peace is quite different from the situation in the past. In the nineteenth century, even excluding the period of the Civil War, the United States averaged about one conflagration per year—conflagration defined in this case as a fire involving groups of buildings that destroyed property valued at the time at $1 million or more.[1] No part of the nation was exempt: great fires incinerated parts of city centers from Portland, Maine, to Charleston, South Carolina, downtown Seattle as well as Chicago.

As bad as the nineteenth century had been, the fires in the first decade of the twentieth century portended, if anything, that the conflagration problem would get worse. In the first four years, fires destroyed large sections of Jacksonville, Florida; Paterson, New Jersey; and Rochester, New York. On February 7, 1904, eighty blocks of Baltimore's business center were consumed in one of the worst fires in the nation's history. In one horrible day, Baltimoreans lost property valued at roughly $50 million to fire. Two years later, following the April 18, 1906 earthquake, the most destructive of all conflagrations leveled San Francisco. The estimated loss—resulting principally from the fire rather than the earthquake—amounted to 25,000 buildings and other property worth $350 million. Then, in 1908, two different sections of the small city of Chelsea, Massachusetts, burned, first in April, causing a loss of 3,500 buildings valued at $12 million, and again in September, in a less destructive fire.[2]

But confounding expectations, the burning did not continue. This time the letup was not a cyclical variation but the beginning of a new long-term trend. Instead of burning through urban centers, the conflagrations and multibuilding fires in the next few decades tended to strike in smaller cities and remote locations, or, if in big cities, in industrial districts rather than the downtowns. Yet even in these places, fires occurred less often and did less damage. The overall trend pointed downward. In the 1980s and 1990s, even multibuilding fires have been rare. Today, certain kinds of places remain vulnerable to sweeping fires, such as older neighborhoods of closely spaced, wooden homes, especially those with businesses mixed in; older industrial areas; and the borders of suburbs and wilderness, where forest fires can enter the city. Arson and civil disturbance are unpredictable factors and have been the source of neighborhood devastation within recent memory. But today we need not fear that a fire in a downtown building will get out of control, spread, and ravage the city, as happened often in the past.[3]

The average person assumes that the reason for this turnabout must have to do with better fire fighting, and certainly the characteristics of a community's fire-fighting service and water supply matter, but they are not the whole story.[4] What matters even more are the characteristics of a city's building stock. Buildings can be constructed so as to contain and withstand a fire, or to serve as fuel. When a building cannot contain a fire, and the fire spreads to neighboring buildings, because these cannot fend off the attack, then the seeds of a conflagration are sown. The effectiveness of fire fighting is much diminished when a blaze involves many buildings. Big fires in the past often created their own gales, which fanned the flames and sent brands flying, starting fires far from where they started. Superheated gases ignited distant buildings before the flames had even reached them. For this reason, regardless of how a fire starts, poor quality construction is a factor in propagating it.[5] Therefore, a defense against this spread is to construct buildings so that a fire burning inside cannot break out, and one on the outside cannot break in. Today, buildings that perform in this way are called fire-resistant or fire-resistive.

A number of characteristics contribute to making a building fire-resistant. First among these is the materials used to build it, which must be noncombustible and able to hold a fire inside long enough to allow firefighters to extinguish it. Another feature of a fire-resistant building is "compartmentation," which involves subdividing the inside with fire-resistant floors, partitions, enclosures around shafts, and so on, to create a series of barriers that will confine a fire to the room or floor where it starts, or at least slow its spread through the building. To prevent fires outside a building from breaking in, buildings are covered with noncombustible roofs. Although window protection is now obsolete, in the past, windows often were built with fire-resistant shutters, wire glass glazing, and noncombustible frames. Finally, the best-protected buildings contain automatic fire-extinguishing devices (sprinkler systems) and an internal water supply for fire fighting (standpipes with hose outlets). A city with many fire-resistive buildings is less likely to burn down than one filled with ordinary, combustible buildings.

Proof that the characteristics of a city's structures matter is offered by Western Europe's superior fire record in the nineteenth century. In major European cities, buildings typically had heavy masonry walls, parapets at party walls, noncombustible roof coverings, and sometimes, notably in Paris, noncombustible floors. Fires that broke out in such buildings usually stayed inside, where they could be extinguished more readily. Thus, the majority of all conflagrations in Europe and North America between 1815 and 1915 occurred in North America, mainly in the United States. America had more great fires even though large cities in Western Europe were at least as densely settled as those in the United States, and the well-staffed and equipped fire departments in big American cities outclassed those of their European counterparts. Fire safety advocates joked that the reason America had such excellent fire departments was because it had such bad buildings.

This is not to say that the reason America had sweeping fires was because it had combustible buildings. Fires have to start somehow, and the number of building fires overall undoubtedly reflects differences in hazards among the countries, a function of varying methods of cooking, heating, lighting, and so on. Possibly Americans inhabited a more hazardous world than did the Europeans. But even if the Europeans had fewer accidental fires, those that did start were more likely to be contained, whereas fires in America often spread. Rather than burn up their money in a blaze, Europeans invested it up-front in their buildings—a trade-off that spared a community the heartache and disruption of conflagration, and paid back, through losses avoided, in the long run.[6]

That the characteristics of buildings helped fuel or check fires was not a foreign concept to American colonists and immigrants, some of whom came from places that regulated building construction in order to avert conflagration. For example, a Proclamation for 1620 required all new houses within five miles of London's gates to be made of brick or stone. A law enacted for Edinburgh a year later called for tile, slate, or stone roof coverings on future houses. In the eighteenth and early nineteenth centuries, several English towns—significantly, places that had suffered sweeping fires—banned thatch roofs.[7] A number of colonial settlements suffered fires early on, prompting authorities to lay down similar sorts of rules. Typically, the laws regulated chimneys and prohibited builders from using certain combustible materials—wood and thatch—in the exteriors of buildings.

But if people dreaded widespread fires, and at the same time knew that by using noncombustible materials such fires might be avoided, why did they have to be compelled by law to use them? There were several reasons why, the most obvious being the high cost of good materials: in America, noncombustible materials such as brick, stone, slate, tile, and metal cost more than combustible alternatives, such as timber, wood shingles, and thatch. The difference in price between a building with brick walls and a slate roof, and one with clapboard walls and a shake roof, varied from region to region, and narrowed over time. Nevertheless, in the nineteenth century, people always paid a premium for noncombustible materials. Moreover, for many Americans, brick buildings offered no attractions apart from fire resistance. New Englanders, for example, preferred their snug wood houses to ones with cold, damp masonry walls.

Another reason people did not bother to put up substantial buildings was that they could rationally expect to avoid being in the path of a conflagration. Although general fires occurred somewhere in the nation *on average* once a year, they did not strike every year, and predicting when or where they would next strike proved impossible. No one in 1872 imagined that Boston's business center, with its impressive granite buildings, would burn down; but in that year, it did. In the early twentieth century, fire protection authorities believed Lower Manhattan, with new high-rises interspersed among dilapidated warehouses, tenements, and hazardous businesses, would surely burn down; but it did not. Annual fire losses represented a waste of resources for the nation as a whole, but individuals were unwilling to shoulder the burden, in the form of more expensive buildings, to help reduce this loss. Few Americans ever personally experienced a great fire, and this made them inclined to take their chances and build as they pleased. There is no evidence that nineteenth-century Americans in general worried much about fire. The modern attitude toward community protection, "rampant indifference punctuated—in the wake of a building catastrophe—by occasional outrage," prevailed in the past as well.[8]

Third, people could shift the cost of protection from themselves to others. One way they did this was by purchasing fire insurance, which was widely available by the mid-nineteenth century. Instead of constructing an expensive, substantial building, an owner could put up a cheaper, less safe one, but still avoid the ruin a fire might bring by making an annual premium payment. As long as he could replace lost property should his building burn, he saved money by taking this latter course. Another way individuals socialized the risks of their building choices was by not paying for fire-fighting services in relation to the hazard posed by their buildings. When municipalities financed their fire departments through a property tax assessed according to the value of a property, and not according to how likely a property was to require fire-fighting services, then the good properties might end up paying for the bad ones.

A final impediment to getting people to construct better buildings voluntarily was that brick walls were no guarantee against conflagration. This is because the usual brick, stone, or even iron-front building of the nineteenth century was simply a woodpile enclosed in noncombustible walls. Behind its substantial-looking facade, the typical masonry building of the nineteenth century had wood-framed floors; a wood-framed roof covered with boards; wood-framed partition walls covered with wood lath; wood stairways; and wood wainscoting and finishes. On the exterior, around and above their slate roofs, such buildings might have wooden cornices, wooden dormers, wooden gutters, a wooden steeple or cupola, and wooden window and door frames, which could communicate a fire burning outside to the interior. It was this combustible material that burned in a general fire, rendering masonry buildings into heaps, as the great fire of 1872 did to Boston's granite and brick mercantile blocks (Fig. I-1). While marginally more secure than a city with wood-walled buildings, a city with "brick" buildings still contained plenty of fuel. To sum up, from the standpoint of the average owner, constructing a building that would protect against fire loss cost him more; may have been unnecessary, thanks to the unpredictability of fires and the predictability of public fire departments; could be insured against anyway; and could not guarantee security.

Nevertheless, some owners voluntarily undertook to make their buildings safe from fire. These owners sought a greater level of security than that which an externally noncombustible building could provide. It was not that their buildings were especially prone to having fires. In fact, owners of the kinds of buildings that burned most often, such as hotels and theaters, did not fall into this group. The members shared several characteristics. For one, their buildings contained valuable contents, usually property, but sometimes people under the owners' care. In addition, many

Fig. I-1. Ruins of Boston after the great fire of November 1872, at the edge of the burned district. The fire leveled nearly eight hundred buildings. The building in the center was Boston's new fireproof post office, under construction at the time. It survived undamaged. Photograph by J. W. Black, "Ruins of the Great Fire Looking Toward the New Post Office," 1872. Courtesy of the Boston Public Library, Print Department.

owners in this group had suffered devastating fires and built secure buildings as replacements. Third, they tended to be comparatively wealthy or at least could command the resources necessary to pay for the security they desired. Owners with these characteristics comprised the clientele for "fireproof" buildings.[9]

The first fireproof buildings in America appeared at the end of the eighteenth century, and methods of making buildings fireproof developed continuously, with systems of fireproof technology—essentially technological regimes—succeeding one another over the course of the century. Whether built in the vault system of fireproof construction, dating from the late eighteenth century to the 1840s, or in reinforced concrete, introduced at the close of the century, all had one feature in common: the constructive parts were made of materials other than wood. Thus, fireproof buildings were noncombustible inside as well as outside, with noncombustible floors, posts, interior walls, and, by the second half of the century, roof frames. To construct them, builders at first adapted traditional materials and building methods, and later introduced completely new materials, such as iron and steel beams, hollow tile, and concrete. These latter materials became the standard ones for commercial construction in the twentieth century, but were originally developed for fireproof construction, as noncombustible substitutes for wood.

Fireproof buildings came into being as a private response to the inadequacy of public fire control measures and the limitations of fire insurance. Fire laws that prohibited builders from putting up structures with wooden walls were effective at best in the long run. Fire insurance could do little for owners who considered the contents of their properties to be irreplaceable, or who could not accept the sort of disruption a fire would cause. Most of the fireproof buildings were erected by governments, financial services companies, and institutions (e.g., libraries, museums), although wealthy families occasionally built fireproof homes and fireproof high-rise apartment houses popped up in many cities at the end of the nineteenth century. These owners constituted a very small group through most of the nineteenth century, yet they called the materials and means of fireproof construction into being.

By the 1880s, the technology of fireproof construction had become so well established that several cities required developers of certain kinds of buildings—specifically, very tall buildings and theaters—to employ them. By the turn of the century, many large cities mandated fireproof construction for tall buildings. These laws spurred the development of the methods and materials of fireproof construction, and, as long as developers chose to build skyscrapers, led to an increase in the stock of fireproof buildings. As new fireproof buildings replaced old combustible ones in city centers, the risk of conflagration abated. Fireproof construction technology predated any city's decision to mandate it; however, without the laws, its use would not have become sufficiently widespread to have the safety effects that it did.

It is possible to trace the history of fireproof construction because contemporaries recognized fireproof construction as a distinct system and called buildings constructed with these systems "fireproof." The members of the fire protection fraternity of the nineteenth century—the architects, fire underwriters, engineers, and materials manufacturers who worked on fireproof buildings—generally agreed on what constituted best practice fireproof construction in any period. Unfortunately, the public never understood what the experts meant when they called a building "fireproof"; the public might apply the word to buildings with noncombustible exteriors only. Worse, the average person assumed it meant that nothing in a "fireproof" building could burn, confusing the *structure*, which might be fire-resistant, with the *contents*, which rarely was. People were often misled by owners who claimed that their ordinary buildings were fireproof. Fires in hotels advertised as "absolutely fireproof" that were nothing of the kind led people to doubt the value of fireproof construction (Fig. I-2). In order to put an end to the confusion surrounding the word *fireproof,* many in the fire protection community made a concerted, and largely successful, effort to purge it from common speech and substitute the word *fire-resistive.*[10] Today, the word *fireproof,* in its technical sense, is obsolete. Nevertheless, building codes and construction handbooks used it well into the twentieth century.

While the methods of creating them, and expectations for their performance, changed over time, nineteenth- and early-twentieth-century fireproof buildings aimed to be ideal, specifically, to be "capable of withstanding the worst sort of fire regardless of the size or occupancy of the building."[11] Fire-resistive alternatives to fireproof construction also existed in the nineteenth century, the principal ones being semifireproof and slow-burning construction.[12] Introduced in the 1870s and 1880s, these systems contained wood, but protected it with noncombustible covers, or else arranged the timber in a special way so that the structure would not burn as readily as conventional framing. Until fire protection authorities made fire resistance a function of time, these systems represented grades of construction with respect to safety, ranging from fireproof at the top, to semifireproof, slow-burning, ordinary, to frame. Eventually, fire protection experts classified materials and assemblies according to how long they were expected to hold up in standard fire tests. In today's building codes, the most fire-resistive category of construction goes by the unevocative designation, "Type 1."

By the first decade of the twentieth century, fireproof buildings had survived fires that would have destroyed ordinary buildings and proved themselves worthy of their name. Yet this success had an unintended consequence: everyone assumed a fireproof building would be safe for the people inside. Fire protection experts gave little thought to emergency egress in case of fire. The public mistakenly believed that everything in a fireproof building was safe. Thus, whenever people died in a hotel or ten-

Fig. I-2. A cartoon mocking the idea of "fireproof" construction. While fire protection experts generally agreed on what constituted fireproof construction, building owners often misappropriated the term for their combustible properties. Bettmann Archive.

ement fire, the press typically called for more fireproof construction, not for better exits. Undoubtedly because of this confidence in the safety of fireproof buildings, building codes contained no special egress requirements for tall buildings, all of which were fireproof, even though these would be difficult to evacuate quickly.

The technology of the fire-resistive building had advanced the hard way, through lessons taught by conflagrations and building fires, and the road to improved egress was an equally hard one. Although fireproof buildings had been involved in major fires, in nearly all cases they were unoccupied at the time. Few people appreciated the danger occupants faced in a fireproof high-rise with the contents on fire. Then, in 1911, a fire in a tall, fireproof building killed 146 people. The event proved to be a turning point in the history of fire safety practice, after which fire protection authorities made safeguarding human life a central focus of their work.

This study traces the development, and consequences, of fireproof construction in America. It covers the period of its creation through to the point when all the principal systems of fireproof construction had been introduced, from the close of the eighteenth century to the opening of the twentieth.

1

THE SOLID MASONRY
FIREPROOF BUILDING,
1790–1840

Security in Public Buildings can alone be obtained by the total abolition from them of combustible substances, except for the most immaterial parts.

ALFRED BARTHOLOMEW, British architect, 1839

In 1631, the year after the arrival of the first settlers from the Massachusetts Bay Company, colonial authorities passed a law prohibiting wooden chimneys and thatch roofs in Boston. Boston suffered its first Great Fire in 1653, after which authorities took various measures to fight fires. But following a fire in 1676 and numerous fires in 1679, they ruled that only brick and stone buildings could be constructed in Boston. In a similar effort to avert general fires, the governor and Council in New Amsterdam appointed firemasters to inspect chimneys, and in 1656, the governor ordered residents to remove thatch roofs and wooden chimneys.[1] The purpose of these early building regulations was not so much to control fire in individual buildings as to prevent sweeping fires that endangered the whole community. The laws regulated the materials on the *exterior* of a building—typically banning wood walls and thatch roofs. Building codes, which today govern every facet of building construction in municipalities across the United States, descended from these early fire laws.

By the early nineteenth century, most large American cities had fire laws. Unlike most other laws, building rules were intended to prevent problems rather than punish offenders after the fact. Typically, they defined a district—what came to be called the "building limits" or "fire limits"—and prescribed construction materials and the minimum thickness of walls for buildings within the district. The laws could cover other matters as well, such as prohibiting hazardous behavior (smoking in barns or carrying open lights in the streets) and regulating hazardous trades.[2] Although they principally concerned buildings, since these were fire laws, fire officials—the old firemasters and fire wardens, later municipal fire department officers—usually enforced them.

Despite their good intentions, the laws did not prevent conflagrations from occurring, mainly because they controlled only the materials of the exterior, which left much inside the structure to burn. Yet finding a substitute for a material as cheap and strong as wood for interior framing was no easy matter. Some early efforts at structural fire protection covered wood with some sort of noncombustible material, such as thin iron plates or mixtures of mortar and other ingredients placed over wood floor frames or between joists, to insulate the wood.[3] These products performed well in experiments; whether they worked in actual fires is unknown. There was the danger that wrapping wood too thoroughly, so as to cut off all air, could cause it to rot. No inventor succeeded in creating a great desideratum, fireproof paint, although many tried.

Some people applied their ingenuity to the problem of creating a building without using any wood structurally. Avoiding combustible interior finishes was trivial: you could substitute plaster wainscoting and decorative molding for wood trim; tile floor coverings instead of boards; and brick partitions instead of frame and lath. The real challenge, however, was finding noncombustible alternatives for the spanning parts of the structure, the floors and roof, in order to create horizontal barriers through which a fire could not pass. In the eighteenth century, some designers and property owners began to experiment with adapting masonry vaults to construct the spanning parts of buildings. The solid masonry, or vaulted, buildings they created were the first kind to be called "fireproof."

Vault construction has been used since ancient times, for decorative effect (as in the ceilings of religious buildings) as well as for permanence and security. Load-bearing vaulted spans, whether in buildings or bridges, usually were in the form of barrel and intersecting, or groin, vaults (Fig. 1-1). To build these vaults, first a temporary form, called a "center," would be set up; then masons would lay the arch. For a bearing span, the top of the arch would be filled in to make a level surface for walking. Vaulted spans inevitably weighed a great deal and required substantial abutments—walls and piers—to support them and contain the thrust of the arch. Thus, a building with vaulted floors had to have thicker walls than one of a similar size with timber-framed floors, which added to its cost. The temporary "centers" also represented an extra item not required in timber construction.[4] Moreover, only skilled masons could build vaults, especially the complicated groin type. Because of the expense, before the eighteenth century masonry vaults mainly appeared in important buildings, such as churches, fancy homes, and public buildings. The new idea that emerged in modern Europe was to apply vault construction to ordinary buildings in order to make them fire-resistant.

Fireproof buildings made of masonry vaults first appeared in Great Britain in the middle of the eighteenth century. At this time, the English architectural writer Batty Langley proposed making "brick floors, with arches, groined, or coved ceilings," in order to "prevent the sad consequences of fire in dwelling-houses." He recommended limiting wood generally: a builder might finish floors with plaster, make wall trim out of stucco and staircases of stone. A vaulted ceiling under the roof, Langley

Fig. 1-1. Barrel (*left*) and groin (*right*) vaults. *Below,* William Strickland's 1832 design for a groin arch floor. The section through one of the supporting walls shows the method of construction. The black lines at the springing of the arch are the ends of an iron tie-rod, not shown on the drawing. William Strickland, "Girard College, Competition Entry, Detail of Classroom, 1832." The Athenaeum of Philadelphia.

believed, would protect the roof and make the building altogether safe from fire.[5] He recommended this type of construction for the country residences of the nobility, which were difficult to protect, should a fire break out, because of their remoteness. But he listed no examples of such houses, so perhaps this was a suggestion, not a description of actual practice. Langley apparently did not know that a few vaulted fireproof buildings existed in England at the time he was writing: several shops at the royal dockyards, used for hazardous manufacturing processes such as for melting pitch, had roofs made of masonry vaults.[6]

Also around this time, a retired French army officer proposed another way of adapting masonry construction to make buildings fire-resistive. Count Felix-Francois d'Espie based his system on the "flat arch"—a kind found in the Roussillon and Languedoc regions of France and areas bordering these in Spain. The arch was made of large, thin tiles joined with a fast-drying sort of plaster. He described both the system and his own house in Toulouse, built according to his ideas, in a small book that came to the attention of a British architect. Translated into English to encourage transfer of the system to England, it appeared around 1756 under the title, *The Manner of Securing All Sorts of Buildings from Fire.*[7] However, the system proved difficult to transport successfully. English masons, not trained in flat-arch construction, could not execute it, and the special mortar and tiles, although plentiful in certain parts of France, were, as far as contemporaries knew, not produced in England. Although d'Espie's idea faded quickly, a century later, the Spanish architect Rafael Guastavino introduced his own adaptation, this time with success. Guastavino's construction company built these vaults in many American fireproof buildings, from the late nineteenth through the mid-twentieth centuries.

At the end of the eighteenth century, a few British architects applied vault construction to larger buildings. John Soane, the most prominent of these, first used vault construction in the Bank of England's Stock Office (c. 1792). He ordered the building's domed and arched ceiling to be made entirely of brick and hollow clay pots (a material later called "hollow tile"). As architect for the Bank of England, he designed several more fireproof buildings for the Bank. David Laing, a student of Soane, designed another large, vaulted building: the new customhouse in London (1812–17). Because fire had destroyed earlier ones, Laing made some of the floors vaulted, including the floor over the entire basement level. Above this, he built an enormous vaulted room, the King's Warehouse, which he covered with a spectacular-looking series of "diagonal elliptical ribbed arches, intersected by parabolic vaultings," supported on granite columns shaped like tree trunks.[8]

Yet the inherent technical difficulties and cost of vault construction limited its appeal in Britain. According to a contemporary engineer, writing in 1819, vault buildings were "very uncommon, very expensive, and the principle upon which they are constructed is not at all adapted for the common purposes of life."[9] A few vaulted floors collapsed, including the one above the King's Warehouse in 1825. Moreover, by the turn of the nineteenth century, English designers had a more flexible system of fireproof construction available to them: the iron and brick arch system. Vaulted floors continued to be built in Britain in warehouse cellars, for fire-safe storage, well into the nineteenth century.[10] In the last quarter of the eighteenth century, when British designers started experimenting with structural iron, the masonry vault system made its way to America. American architects used the system to a much greater extent—for larger buildings and over a longer period—than the British ever had.

The Early Days of Vault Construction in America

America's timber wealth made this country an inhospitable place for masonry construction.[11] Few colonial buildings contained vaults of any kind, even in cellars. The frames inside all of the large buildings of the early national period—the north wing of the U.S. Capitol in Washington, the statehouse (Independence Hall) in Philadelphia, and the second statehouse in Boston, for example—were made of timber. Nevertheless, in the first half of the nineteenth century, perhaps a few dozen vaulted buildings were erected in the United States. The vault system defined what it meant to build fireproof: if someone wanted a fireproof room or building before the 1840s, they built it of solid masonry construction.

The system had its start in Philadelphia, when the city decided to replace its dilapidated jail with a new one, the Walnut Street Goal, which became America's first fireproof building. Construction of the jail began in 1774, and it took about ten years to complete. More impressive even than its imposing stone facade was the groined and barrel-vaulted floors inside. "There are eight rooms upon each floor," a contemporary wrote in 1798, "all groin arched for the two fold purpose of *securing against fire and escapes.*"[12] The building went up under the superintendence of the important Philadelphia carpenter-architect Robert Smith, who learned the building trade in Scotland before immigrating to America as a young adult.[13] Whether Smith, a carpenter, designed this masonry jail is unknown. Carpenters always worked on vault buildings: they made the wooden centers for arches and vaults. However, none of Smith's other projects had vaults and, since it was common practice at the time for one architect to furnish plans and another to oversee the construction, someone other than Smith may have designed the building. At any rate, Smith died before it was complete, so another architect finished the jail—perhaps a recent British immigrant with experience building vaults. The architect of a fireproof prison block erected around 1790 behind the Walnut Street jail likewise is unknown. The year before, Pennsylvania had enacted a new penal code that called for hard labor instead of corporal or capital punishment for many crimes. Lawmakers ordered a prison with solitary cells—the first of its kind—to hold recalcitrant prisoners, and required that the building be made of brick and stone "upon such plans as will best prevent danger from fire."[14] The modern penitentiary descended from this building.

After these two precocious examples, the history of vault buildings truly gets under way, following the arrival in America of the English architect and engineer Benjamin Henry Latrobe. An experienced designer, Latrobe brought knowledge of vault construction with him. He helped propagate the system in two ways: first, by using it in high-profile projects that served as models, and second, as a teacher of architects who became important practitioners and advocates of the system.

Latrobe landed in Norfolk, Virginia, in 1796 and offered the governor his services to design a new state penitentiary at Richmond. Virginia's legislature, following Pennsylvania's example, had just enacted a penal code that imposed imprisonment and labor, rather than capital punishment, for all felonies except murder. The legislature wanted a prison like Philadelphia's, including a separate "penitentiary house" in the prison yard, which should provide light, air, and security from fire. Latrobe had not seen this model, however, and his 1797 plan drew on other sources. It called for a horseshoe-shaped building closed across the ends with another block of buildings, creating a half-circle prison yard within. The buildings contained a mix of cell types along with offices, kitchens, baths, workshops, and so on. Like the Philadelphia prison, many of the cells in the Virginia State Penitentiary were vaulted. Difficulties with this project, disputes over his salary, and the prospect of a better job prompted Latrobe to travel to Philadelphia, partly to study the Walnut Street jail and partly to scout for new design work. He succeeded in obtaining a commission to design a building for the Bank of Pennsylvania and moved to Philadelphia to superintend its construction.[15]

The Bank of Pennsylvania (1798–1801) became the first fireproof building erected by private interests rather than by government. Latrobe modeled its plan on that of John Soane's vaulted Bank of England Stock Office of 1792: both were rectangular buildings containing a large, central room flanked on either side by smaller rooms. This three-part plan became a popular one for vaulted bank buildings in the United States. In the Pennsylvania Bank, Latrobe finished the center room—which served as the banking hall—as a cylinder and covered it with a relatively flat dome, lighted at the top by a lantern. Few domed buildings existed in the United States at this time and in all other cases, the domes were framed in wood. Latrobe made the bank's dome entirely of masonry: brick covered with marble slabs (Fig. 1-2).

While it represented a great technical accomplishment, the dome, a feature of Roman architecture, did not fit historically with the bank's Greek-style porch. Classical Greek temples had pitched roofs framed in wood. Working in the Greek style, the architectural fashion of the early national period, therefore posed a particular problem for architects attempting to reproduce classical models while avoiding wood. Faced with few alternatives, they often resorted to historically inaccurate domes in order to give their Greek-style buildings noncombustible covers. Contemporary architectural writers criticized the architects of these hybrid buildings as poor students of the classical model, but this was unfair: the roots of the design choices were more practical than aesthetic.[16] Several architects did succeed in creating pitched roofs using only brick and stone, but these instances were very rare. More often, because of the cost and difficulty of making a masonry roof of any sort, architects compromised and put timber roof frames on vaulted buildings. They sometimes protected the frame by installing a vaulted ceiling underneath it.

Latrobe went on to build several notable fireproof buildings for the federal government. In 1803, President Thomas Jefferson, who knew of Latrobe's designs for the Virginia penitentiary, appointed him surveyor of public buildings. Two of his assignments involved completing buildings designed by other architects: a wing for the Treasury Department and the south wing of the U.S. Capitol. Latrobe fitted vaulted structures within walls planned to enclose ordinary construction. He reconstructed the basement of the Capitol with vaults and placed a masonry arch over the new

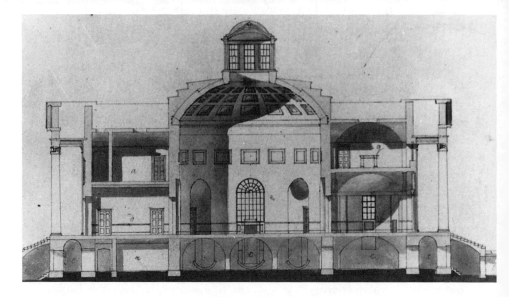

Fig. 1-2. Longitudinal section through the Bank of Pennsylvania (1798–1801) in Philadelphia. The three-part plan of this building—small rooms flanking either end of a large, domed or vaulted center room—became a standard one for vaulted buildings. Drawn by Benjamin Latrobe before the building was completed, the sketch may not represent the bank's final form. Benjamin H. Latrobe, "Pennsylvania Bank, Section from East to West," The Historical Society of Pennsylvania, Philadelphia (Bd 615 L 354 p. 27).

Treasury wing to make it fireproof. The latter project, undertaken in 1805–6, illustrates the main factors that motivated owners to invest in fireproof construction: the Treasury had caught fire in 1801, it contained valuable papers, and the federal government could afford to put up a secure structure. Another of Latrobe's fireproof federal buildings was the short-lived customhouse in New Orleans (1807–8). Although a small structure, it still proved too heavy for the city's marshy soil, and it settled unevenly, causing the walls to crack. With repairs postponed by the War of 1812, the building fell into ruin, and it had to be replaced in 1819, only ten years after it opened.[17]

Latrobe continued to work on private commissions in this period, one of which was a building for the Bank of Philadelphia (1807–8)—probably the second fireproof building to be erected by private interests. None of Latrobe's drawings of the interior have survived, so its construction is a matter of conjecture, but the center room was vaulted. Architect Robert

Mills, who supervised its construction for Latrobe, wrote that he found erecting the building a challenge because of the "novel forms of the vaultings and great span of arches in the center hall, all of which were built of solid masonry and made fireproof."[18]

Latrobe continued designing buildings for the federal government, after leaving Washington at the start of the War of 1812, and then following the war, when he was asked to reconstruct federal buildings destroyed by the British in 1814. While living in Pittsburgh during the war, Latrobe designed several buildings for the new federal arsenal there, the largest of which was a three-story, vaulted arsenal building (1814) (Fig. 1-3). No doubt because of the high cost of executing such a monumental structure, the government abandoned Latrobe's design. Only one of Latrobe's designs for the arsenal, for a fireproof powder magazine—a one-story brick building—was built according to his plans.[19] Powder magazines, for storing gunpowder and ammunition, were among the first buildings in England to be built with vaulted roofs, for fire protection. On his return to Washington, Latrobe must have felt great satisfaction to find that both his fireproof sections of the Capitol and fireproof addition to the Treasury building had survived fires set by the British. But he soon resigned, in 1817, amid charges of extravagance—the sort of accusations made against practically every architect who did work for the government, whether justified or not.

Unlike most American architects at this time, who apprenticed as carpenters and masons in their youth, Latrobe had no background in the building trades. How he learned to design and supervise the construction of vaults, therefore, is uncertain. He explained that he picked up practical knowledge from workmen on the job, by observing "the methods, the tricks, and the knacks of workmen with the technical vocabulary belonging to them."[20] His lack of building experience may explain why his vaults occasionally failed. A vaulted loggia that connected his Treasury Building wing with the president's house collapsed after the centers had been removed. When informed of the disaster, Latrobe wrote to his clerk of works, John Lenthall, "I am very sorry the arches have fallen, both on account of the expense & the disgrace of the thing. But I have had such accidents before, and on a larger scale, & must therefore grin & bear it." In 1808, the vaults over the Supreme Court chamber in the Capitol collapsed, also following the striking of the centers, killing Lenthall.[21] The novelty of the system, the great weight and force of the vaults, made vault buildings more prone to failure than traditional ones.

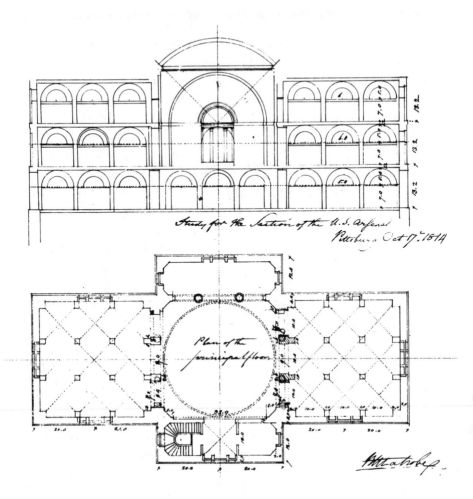

Nevertheless, the vault system spread in the United States, in large part through Latrobe's students. One of these, Robert Mills, became the most prolific member of the second generation of vault-building architects. Whereas Latrobe considered himself an artist whose buildings incidentally were fireproof, Mills advocated fireproof construction as such. Through his writing, he defined the meaning of a "fireproof building" for the public. Mills came to Washington from South Carolina, intending to study architecture, and his friend Thomas Jefferson referred him to Latrobe. During the time he worked in Latrobe's office, from 1803 to about 1809, Latrobe had several fireproof building projects. Mills supervised the construction of one of these, the Bank of Philadelphia, which he considered his first fireproof building. Soon after setting out on his own, he de-

Fig. 1-3. *Above:* facade elevation of Latrobe's design for an arsenal building at Pittsburgh (1814), showing the often forbidding architectural results of the vault system. *Facing page:* section through the building (*top*) and plan of the main floor (*bottom*). The hatched lines in the plan indicate the vaults above, springing from wall to pier. This building was not constructed. Benjamin H. Latrobe, 1814. Pittsburgh Arsenal. Library of Congress, Prints and Photographs Division.

signed a fireproof jail in Burlington County, New Jersey (1808). Like the penitentiary block at the Walnut Street jail and Latrobe's Virginia prison, Mills designed his prison to have vaulted cells and he urged Burlington officials to adopt the noncombustible materials. "*Humanity* as well as *Interest*," he wrote, "plead in favor of this, when we refer to the many melancholy Instances on record of persons confined, perishing in the flames from inattention or being forgotten during the Confusion."[22]

Another of Mills's early projects was a design for fireproof wings for the Pennsylvania statehouse (now Independence Hall) in Philadelphia. Built around 1812, these wings on the east and west sides of the historic building housed courtrooms as well as government offices containing irreplaceable public documents. Another reason for making them fireproof was so they would pose no hazard to the cherished statehouse. Only two stories tall, made of brick, these modest additions resembled houses more than the banks, prisons, and federal buildings that comprised America's small stock of fireproof structures to date. Mills subdivided the interior of the basements and first floors with bearing walls, and covered the rooms with barrel and groin vaults. The narrow, dimly lit rooms that resulted, with their curving ceilings, bore an unfortunate resemblance to caves. The second floors of both wings, which held the courtrooms, were not vaulted and had much better light. The roof was framed in wood.[23]

In the 1820s, Mills served as architect and engineer for the state of South Carolina and designed many county courthouses, jails, and other structures for the state, which he built "as far as was practicable" in "the fireproof system."[24] One of his assignments was a hospital for the mentally ill in Columbia, the state capital. Again, Mills urged officials to invest in a "thorough vaulted fire-proof structure . . . in consequence of the helpless character of its inmates." His plan was accepted and the original state hospital (1821–27) was vaulted. It also featured iron window sashes and shutters to resist fire and bar escapes. Mills's best known building, the vaulted Record Office in Charleston, went up in this period. Popularly known as the Fireproof Building, it also contained iron window frames, sashes, and shutters.[25] All of Mills's county courthouses had vaulted first floors. However, they also had timber-framed roofs, a concession to economy.

Before the 1830s, the federal government tried to spend as little as possible on its buildings, even if this meant housing government offices in combustible structures. For the most part, it bought and adapted existing buildings for its purposes. Less commonly, it built new buildings, sometimes ordering these to be fireproof or at least partly fireproof. Thus, several new customhouses and federal buildings in Washington were fireproof. When Congress authorized new customhouses for four New England cities, it did not specify the type of construction. Robert Mills, who won the commission to design these four buildings, naturally recommended making them fireproof. He estimated that making them fireproof would add just $1,000 to the cost of completing each building "in the common way," representing a 6.7 to 10.5 percent increase, depending on the structure's size. This modest price difference helped convince the Treasury Department to approve Mills's fireproof plans. When these four customhouses—at Middletown and New London, Connecticut, and New Bedford and Newburyport, Massachusetts—went up in the early 1830s, they became the only large, vaulted structures in New England. Like his South Carolina Record Office, all had vaulted basements and first floors; iron window sashes and frames; and iron doorframes. Mills supplied the designs and specifications for the buildings, but did not superintend their construction and had no control over the final cost. As it turned out, he underestimated the actual cost, which came to 12 to 38 percent more than his estimates.[26]

William Strickland, another of Latrobe's students, also became an expert in fireproof construction. Strickland's father had worked as a carpenter on Latrobe's Bank of Pennsylvania and apprenticed William, who had already learned bricklaying, to Latrobe. In Latrobe's office, he learned to draw and design, and during the War of 1812, he gained executive experience by overseeing the construction of fortifications around Philadelphia. Thus prepared, Strickland set up an architecture practice in Philadelphia after the war; his commissions included fireproof building projects for government as well as private clients.[27]

In 1818, Strickland won an important competition to design the Second Bank of the United States in Philadelphia (1819–24). The bank's president, the Grecophile Nicholas Biddle, wanted the building to be in the Greek style and fireproof. Strickland planned the bank to have three sections, like Latrobe's Pennsylvania Bank, although he covered the central banking room with a barrel vault rather than a dome. He may have modeled this room on the barrel-vaulted one in the First Bank of the United States, completed two decades earlier and located around the corner from the Second Bank. The First Bank was not thoroughly fireproof, while the Second Bank was, with the exception of its historically accurate, pitched roof, which was framed in timber.[28]

Another of Strickland's vaulted buildings was a residence for retired naval personnel, the United States Naval Asylum (1827–33), later called the Naval Home, in Philadelphia. Like state mental hospitals and prisons, this kind of residential structure called for built-in fire protection because occupants might be unable to escape quickly in an emergency. Strickland described the building he planned in an 1826 letter to the Secretary of the Navy, explaining that it would contain "about 250 dormitories and will be arched throughout making the whole building fireproof to the roof."[29] As built, the structure was 385 feet long and three stories above ground, consisting of a central block with wings. A masonry dome covered the center room, supported on pendentive arches; the room served as a chapel. The wings, which contained the bedrooms, were lined with balconies. At the second and third levels, cast iron posts supported the balconies, making this one of the first buildings in the United States to have cast iron columns. Strickland undoubtedly specified iron rather than wood posts in the balconies for the sake of fire resistance. He installed iron instead of wood in balconies and stairways in several other projects in late 1820s through the early 1830s: three mints for the Treasury Department and an almshouse for Blockley Township in Pennsylvania, all fireproof.[30]

Surprisingly, no significant fireproof construction tradition developed in New England. The presence of Latrobe, Mills, and Strickland in Philadelphia accounts for the comparatively large number of fireproof buildings there. Philadelphia landowners could see the local vault buildings and this must have helped sell the system. Mills's four customhouses represented the only vault buildings in New England, and the lack of examples of the type may have discouraged owners from adopting the system. British architects who settled in Boston, who may have had experience building masonry vaults, received no commissions to design them.

Nevertheless, the region contained a few partly vaulted buildings, which were the work of native New Englanders who had no direct connection with Latrobe. Perhaps the first such building was the customhouse in Boston (c. 1810), designed by the real estate developer Uriah Cotting. An 1828 guidebook described the building as having a vaulted ground floor, with the arches springing from bearing walls; floors paved with stone; and a stone stairway with an iron railing.[31] Another partly vaulted Boston building was Massachusetts General Hospital, designed in 1818 by Charles Bulfinch and completed by Alexander Parris. With its granite facade and central staircase with cantilevered granite treads, the hospital was an early example of granite architecture, which became a distinctive feature of Boston. Parris made parts of the cellar and basement levels, as well as the corridors on the principal level, vaulted. Elsewhere he used timber. The stone paving in the main entry and the granite staircase were sufficiently unusual to be noted in an 1838 Boston guidebook, but contemporaries did not describe the hospital as fireproof.[32]

New Englanders seemed to prefer the approach of applying vaults only to part of a building. Two more examples of this were the Boston branch of the Bank of the United States and a new Suffolk County courthouse, also in Boston. Construction on Branch Bank began in 1824, after the nearby Massachusetts General Hospital with its vaulted cellars had been completed. Designed by Solomon Willard in the Greek style favored by bank's president, Nicholas Biddle, it had the familiar three-part plan, with small rooms at each end and a large, domed banking room in the center. A contemporary writer described the bank as being constructed with "refractory material (Chelmsford granite)," but did not describe the interior or call the building fireproof.[33] The new courthouse for Suffolk County (1832–36) contained fireproof rooms for storing county records, as state law required. Alexander Parris, superintending architect on this project, introduced a novel, "spheroidal form" of vaulting to cover the two fireproof rooms. These small domes, he argued, exerted less thrust than groin arches.[34]

The only thoroughly vaulted buildings in the region before 1837, apart from Mills's four customhouses, were gunpowder magazines. British vault-covered magazines, such as the ones built by the Board of Ordnance at navy yards, served as models for the American military magazines.[35] Latrobe designed several fireproof magazines for the federal government, and the one for the federal arsenal at Pittsburgh did go forward. America's woeful military readiness, exposed by the War of 1812, prompted the Army to select sites around the country for storing weapons and ammunition. One of these sites was at Watertown, Massachusetts, and the Army hired Parris as architect for the new arsenal structures. Its massive stone magazine, built in 1817, was probably his design. This rectangular building, about 80 feet long and 32 feet wide, had a line of columns down the center that supported groin arches. Although very costly (about $25,000), the danger of an explosion justified the expense: shortly before it was built, an explosion at the arsenal had killed five men.[36] Parris designed several more magazines. In 1836, he built one for the Navy Department at Chelsea, Massachusetts, which had a roof made of small domes like the ones he built in the Suffolk courthouse. Parris built a second fireproof powder magazine at Chelsea, and later, two stone magazines at the navy yard in Portsmouth, New Hampshire. Yet none of Parris's other important projects—the market building and blocks of stores behind Faneuil Hall in Boston, or the naval hospital in Chelsea—were fireproof.

Since they did not have Latrobe to teach them, how did New England architects learn to design vaults? The answer is, from books and observation. New England's vault architects all had backgrounds in the building trades and could apply ideas they read about or saw. The careers of Parris and Willard, and another New England architect, Ithiel Town, show how important books were as a means of spreading construction technology. Parris had apprenticed as a carpenter, and he worked as a builder and designer before the War of 1812. Willard worked as a housewright, carver, and architectural model maker. Since he carved the wooden spread eagle that topped Boston's 1810 customhouse, he would have known its vaults and possibly watched them being constructed. Both men visited the U.S. Capitol and would have seen Latrobe's vaults there—Parris in 1812, when Latrobe personally showed him around, and Willard in 1810 and again in 1818. Willard also worked in Baltimore from 1817 to 1818 and had met Robert Mills.[37] In addition, both were members of the Architectural Library in Boston, the Athenaeum, and the Boston Mechanics' Institution, where they had access to architectural and technical books. Parris amassed his own library of engineering and architectural books.[38] Willard's biographer captured his subject's engineering aptitude when he described him as someone who "always worked with his head as well as hands," and liked to figure out ways to do projects "in some easier, quicker or cheaper way, or in a more perfect manner."[39] Ithiel Town, born in Connecticut, learned the carpenter trade while living in the Boston area and then became an architect and builder. In addition to designing several fireproof buildings, including a new state capitol for North Carolina (1833–40), he patented a bridge truss that was widely used and made him a wealthy man. He acquired one of the largest private libraries in the United States in his day, which included many architectural books.[40]

This first generation of structurally experimental New Englanders taught the next. Isaiah Rogers, who had apprenticed as a housewright as a youth, worked with Solomon Willard in the 1820s before setting up his own design practice. Richard Upjohn and Gridley James Fox Bryant worked with Alexander Parris: the former assisted with the Suffolk County courthouse and the latter, son of master mason and inventor Gridley Bryant, worked with Parris at the Charlestown Navy Yard. This second generation continued the development of fireproof construction.

The Heyday of Vault
Construction: The 1830s

In the latter part of the 1830s, the federal government increasingly opted for fireproof construction for its new buildings, despite the greater cost and longer construction times involved. The trend began after an arson fire destroyed part of the Treasury Department building (but not Latrobe's fireproof wing) in 1833. Congress directed the president to replace the Treasury's burned section with a "fire-proof building."[41] Worried about the insecurity of the building housing the Patent Office, the chairman of the House Committee on Patents introduced a bill "for erecting a fire-proof building for the Patent Office," but it died. Tragically, on December 15, 1836, the old hotel that served as the home for the Patent Office, along with patent records and models inside, burned down.[42] This building also contained Washington's post office. Congress voted funds to build new structures for both offices and ordered them to be fireproof. President Andrew Jackson appointed his friend Robert Mills to construct the new Treasury building, and work got under way in 1836. Mills was also appointed superintending architect for the new Patent Office and later, architect for the new post office, on which construction began in 1839. Mills put vaulted interiors in all three buildings. In addition, all had timber-framed roofs.[43]

The most monumental federal buildings outside Washington were the new customhouses in New York City and Boston. Ithiel Town's 1833 Greek temple design for New York City's customhouse provided so little space for business purposes that the building committee asked Town to revise the plan, but he chose to quit the project. It took eight years (1834–42) and two superintending architects, Samuel Thomson and John Frazee, to complete it. They retained the domed rotunda in Town's plan but placed the dome under a pitched roof, achieving thereby a historically accurate appearance on the outside. At the close of the century, the architect Richard Upjohn wrote an appreciation of this customhouse in which he noted the masonry dome over the rotunda, and the complete absence of wood in the floors and ceilings; even the roof was made "entirely of marble, supported on the groined arches." Such construction, he concluded, made the whole building "invulnerable to fire and enduring as the ages."[44] Not to be outdone by New York, Boston merchants lobbied successfully for a new and equally monumental customhouse, which got under way a few years later. Designed by Ammi Burnham Young, this massive, vaulted structure took even longer to build than New York's—over ten years (1837–49). Young had apprenticed in the building trades and superintended the construction of Vermont's granite, although not fireproof, statehouse in Montpelier (completed 1838). Like New York's, Boston's customhouse was made entirely of masonry.

Although federal and state governments were the main clients for fireproof buildings, some private owners put up notable fireproof buildings in the 1830s. The most prominent of these—and at a cost of nearly $1.5 million, also one of the most expensive—was Girard College in Philadelphia (1832–48). The Philadelphia merchant and philanthropist Stephen Girard left the bulk of his fortune, about $6 million, to establish a boarding school for orphan children. A detail-oriented man, Girard left instructions in his will about how the principal classroom building should be built. He called for a three-story building, each floor to have four main rooms, 50 feet square. The ceilings over the rooms were to be arched, 50 feet in the clear, with vaults starting 15 feet above the floor. Girard also required that the building be "fire proof inside and outside," meaning, "The floors and the roof to be formed of solid materials, on arches turned on proper centres, so that no wood may be used, except for doors, windows, and shutters."[45]

While Girard undoubtedly intended the buildings to be plain, he did not specify an architectural style, which allowed his executors to select something grand. The Philadelphia City Council held a competition and chose a classically inspired design by a young architect, Thomas U. Walter. The son of a mason, Walter had worked with his father on the Second Bank of the United States. He learned drawing and design in the office of bank's architect, William Strickland, and thus brought the influence of Latrobe to a third generation.[46] While accepting the Council's architect, the president of the college's trustees and head of the building committee, banker Nicholas Biddle, rejected the winning design. He wanted a peristyle Greek temple—one surrounded by freestanding columns—and therefore Walter made a new design intended to satisfy Biddle and fulfill Girard's requirements. As constructed, the classroom building, now called Founder's Hall, had three stories and a basement, and the arched classrooms that Girard described. Walter gave the building a historically accurate pitched roof by constructing brick arches over the domed and sky-lit top floor, which carried 4-foot-wide stone roof tiles, a solution similar to the one in New York City's roughly contemporaneous customhouse. Among the remarkable features of Founder's Hall were cantilevered stone stairways in the entries that wound 50 feet up to the third floor and a vaulted tunnel running under the stepped platform that surrounded the building (Fig. 1-4).

In approving the plans for the college, the trustees committed themselves to spending a great deal of someone else's money. Businessmen were thriftier with their own money, but they did sometimes splurge on buildings, in particular, the "exchanges" where they met to conduct business. Merchants' exchanges first appeared in the early nineteenth century and while they usually contained offices for rent, dining rooms, and sometimes a hotel or library, they always featured a large room in which merchants could meet and make deals. One of Benjamin Latrobe's last projects was a partly fireproof exchange building in Baltimore that he designed with

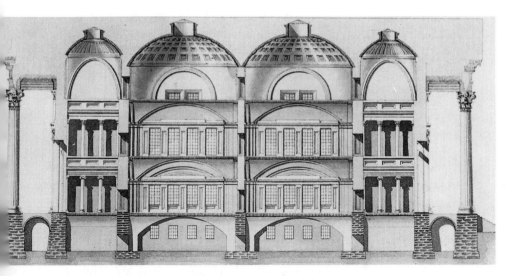

Fig. 1-4. The pinnacle of vault construction: Founders Hall at Girard College (1832–48). Longitudinal section, drawn by Thomas U. Walter in 1834. Thomas U. Walter, "Longitudinal Section of Girard College, 1833." The Athenaeum of Philadelphia.

Maximilien Godefroy (1816–18). In the 1830s, fireproof or partly fireproof exchanges went up in several cities. One of these was the merchants' exchange in Philadelphia, designed by William Strickland and built between 1832 and 1834. It contained a vaulted ground floor and interior partitions made of brick, which served as firebreaks. He made the timber-framed floors fire-resistive by covering them with a layer of mortar, and paved the halls with stone tiles.[47] New York City's businessmen had to build a new exchange when the great conflagration of 1835 swept away their new exchange, which was not fireproof. In contemplating a replacement, New York's businessmen had the example of the fireproof customhouse, which was under construction at the time of the fire and survived unharmed. The projectors of what would be the third exchange (1836–42) resolved that the new building would not suffer the fate of its predecessor and ordered "every room shall be vaulted and rest upon arches, and be made completely fire proof."[48] They gave the design commission to the New England architect, Isaiah Rogers, who was then in New York building the Astor Hotel. The New York City Exchange probably was Rogers's first fireproof building. He designed a magnificent, vaulted building for his clients, which ended up costing nearly $2 million.[49] Diarist Philip Hone, one of the investors, worried that the building would never generate sufficient revenue to repay the debt. As the New York exchange building neared completion, Rogers received a commission to design a fireproof merchants' exchange for Boston (1840–42). According to a Boston guidebook from 1860, the "entire building is fire-proof," with staircases of iron and stone.[50]

Limitations of the Solid Masonry Building

In one important respect, the vault system achieved its purposes: in actual fires, although usually damaged to some extent, vaulted structures remained standing when the flames died down. Latrobe's south wing of the Capitol survived a fire set by the British in 1814 and his 1805–6 fireproof addition to the Treasury building suffered little injury in the 1833 fire that destroyed the rest of the building. An 1877 fire in the Patent Office badly damaged sections of the building, but not Mills's vaulted south wing. The government replaced the upper floor of this wing in the early 1880s out of concern for its future safety, not because it had failed. Lastly, the federal government's circa 1839 United States Appraisers Store (also called Public Store House No. 1) in Baltimore survived the 1904 conflagration, although burning alcohol on the third floor created an inferno that badly damaged the granite columns. After repairs, the building continued in service until 1933, when it fell to a wrecking ball.[51]

But fire resistance could not compensate for the system's obvious disadvantages: besides their high cost, vaulted buildings made poor containers for human activities. The drawbacks stemmed from their unavoidable monumentality. To bear the weight and thrust of the arches, the buildings required massive walls, partitions, and columns, which consumed space on the inside and created deep openings around windows that blocked natural light. The Treasury Department outgrew its customhouse in New York by the 1850s, prompting Treasury Secretary James Guthrie to complain that in no building had the government spent so much to get so little. The acoustics in the vaulted buildings were often bad. Even as he built Founder's Hall for Girard College, Walter warned that the size and materials of the classrooms "render them wholly unfit for use." He advised that "unless a level ceiling is thrown in . . . or some other means adopted to destroy the reverberation, they can never be used for the purposes of school or recitation rooms."[52] The college eventually moved classes to other buildings.

Probably none of the vault buildings, with the exception of the powder magazines, were well suited for their intended purposes. Before electric

lights and central heat, their interiors were dim and uncomfortable. Criticizing Mills's Treasury Department building, the chairman of the House
Committee on Public Buildings and Grounds charged that its central corridor "more nearly resembl[ed] the tunnel of a railroad than . . . a building."[53] Some years later, Abram Hewitt—the prominent iron manufacturer and Democratic politician—recited the drawbacks of vault-style
buildings to contrast them with the alternative system he proposed. He
wrote that "In all these [masonry fireproof] . . . buildings, the basement
story is rendered dark and gloomy and dungeon-like by the immense
number and massive forms of the walls, piers, columns, and groined and
vaulted arches"[54] (Fig. 1-5). All the dead weight in these buildings made
them especially vulnerable to the destructive force of gravity. An engineer
who observed the demolition of Robert Mills's wings to the statehouse in
Philadelphia in 1896 calculated that the vaults weighed an enormous 217
pounds per square foot.[55] Uneven settlement contributed to the failure of
Latrobe's customhouse in New Orleans. When the walls of Mills's customhouse at Newburyport began to bulge out, officials tied them together
with an iron rod.[56]

Not only was the final building less than ideal, but the technical problems associated with constructing vaulted structures often increased their
cost and delayed their completion. Robert Mills wrote that the great masonry span in Latrobe's Philadelphia Bank posed many problems for him,
although he does not say whether this was because of the complicated design or workmen who lacked the skill to execute it. Yet the Philadelphia
region had a large number of masons with experience executing vaults; in
New England and the South, such workmen were scarce. Mills wrote that
his courthouses and jails in South Carolina were built "by mechanics who
were little acquainted with the construction of groin-arches or centering."
In New England, Mills "found not a single bricklayer that knew how to
turn a groin. . . . I had to instruct the workmen, both in forming the centres and cutting the groin."[57] The architect of North Carolina's vaulted
state capitol recruited masons from Philadelphia, some of whom had just
finished working on that city's vaulted Eastern State Penitentiary; he went
as far afield as New York and Boston to find a workforce.[58] In the 1850s,
John Niernsee, architect of South Carolina's capitol building in Columbia,
set up a school to train workers, who then could execute the intersecting
vaults in the building's main corridor.[59]

Unfamiliarity with this sort of work also led contractors to overvalue
the cost of doing it. It was for this reason, Mills believed, that the bids for

Fig. 1-5. Interior of a street-level room in one of the fireproof wings of the Pennsylvania State House (now Independence Hall) in Philadelphia, occupied by the city's Naturalization Office. A photographer documented the buildings around 1895, shortly before their demolition. His photographs capture the gloomy, cave-like quality of the vaulted rooms. A. P. Smith photographs of the fireproof wings of the statehouse, taken circa 1895, in the collection of The Atwater Kent Museum, Philadelphia, catalogue no. 49.20.8.21. Courtesy of The Atwater Kent Museum.

work on the Middletown customhouse exceeded his estimates. "The thorough manner here proposed of giving a Fire proof character to our building," he explained, "is new among our Mechanics, and hence may account for the heavy valuation put by them upon such work."[60] Finally, vault buildings often took an extraordinarily long time to complete. New York's merchants' exchange was under construction for five years; Boston's customhouse took more than ten; and Girard College—although stalled during the early 1840s because of money problems—took almost fifteen years to complete. Businessmen who needed offices, factories, or warehouses could not keep money tied up too long waiting for buildings to be completed.

Mills, for one, tried to overcome one of the drawbacks of the system by attempting to make the walls no thicker than those in an ordinary building. Thus, in the Treasury building in Washington, he used hydraulic cement both as mortar and to level up the arches instead of the usual weak lime mortar. As a fill for arches, the more tenacious mass created by the cement, he believed, exerted less thrust. But the chairman of the House Committee on Public Buildings and Grounds, which oversaw the development of this building, failed to appreciate Mills's innovation. Questioning the stability of the structure and Mills's competence as an architect, the committee invited Mills's professional rivals, Walter and Parris, to evaluate the sufficiency of the walls of this building. Both men produced adverse reports and recommended strengthening the building's walls or changing the system of construction. In his defense, Mills pointed out that he had used the approach successfully, for example, in his state hospital in South Carolina, which had walls only two-and-one-half bricks thick, exclusive of piers. Mills survived this attack and built the Treasury Department according to his own ideas, but no one else adopted them. Despite the dire predictions, all of his Washington buildings still stand today.[61]

Some owners adopted alternative ways to make buildings more fire-resistive, if not completely fireproof. Owners often ordered vaulted cellars or fireproof rooms for their otherwise ordinary buildings. Early examples of this strategy from Philadelphia include the First Bank of the United States (1794–98), which had a vaulted cellar and two small, vaulted rooms on the first floor, and Arch Street Friends Meeting House (1803–5), with its vaulted safe room.[62] In 1812, Massachusetts lawmakers directed every county courthouse to have fireproof rooms for storing public records, and accordingly county officials built new courthouses with vaulted cellars or installed vaulted rooms in existing buildings.[63] Two decades later, the legislature ordered a fireproof addition to the Massachusetts statehouse for storing the Commonwealth's records and papers. Also in Boston, St. Paul's Church (1819–20) had two fireproof rooms under the steps of the portico.[64] In Baltimore, houses on sloping lots often had vaulted cellars; owners built their houses on the high end of the lot and vaults extending under the yards behind. The vaults "formed admirable fireproof receptacles . . . such vaults combined both security and safety against almost all avoidable mischance."[65] This widespread practice of building fireproof rooms in ordinary buildings continued through the first three-quarters of the nineteenth century, at least. Office buildings often contained vaulted rooms that served as safes for tenants.[66]

Another less expensive alternative to vault construction was to subdivide a building with firewalls and to make the floors more fire-resistive, to check a fire's spread. The principal method for reinforcing a wooden floor against fire was to coat it with mortar: filling the space between joists or between an underfloor and the finish surface. Used since the 1820s at least, this technique adapted an old process called "deafening," which employed mortar in a similar way to muffle sound. Although such a floor would burn through eventually, it could remain intact longer than an ordinary floor, giving more time to occupants to escape and firefighters to save the building. Architects used this method to make floors fire-resistant through the third quarter of the nineteenth century and perhaps longer.[67]

Interior brick walls—firewalls—were another important means of containing a fire. An early example of a building with such firewalls was the Wadsworth Atheneum in Hartford (1842–44). Designed by Ithiel Town and Alexander J. Davis, the Atheneum looked like a single building from the outside, but inside was subdivided into three separate sections to house three cultural institutions. A contemporary survey of American libraries described the building as a "spacious, safe and massive edifice," but not fireproof.[68] Similarly, the plan for the old Yale College Library, built 1842–47 and designed by Henry Austin, included firewalls between the central section of the building and its two wings.[69] Ithiel Town applied this principle of subdivision, today called "compartmentation," to his own home in New Haven (c. 1832), which housed his valuable library and art collection. He protected the floors with two inches of mortar and "water cement," and applied plaster directly to the exterior masonry walls, thereby avoiding highly combustible wood lath. A writer in 1839 described Town's house as "substantially fireproof," again indicating that the wood-framed floors—although protected—disqualified it from being called simply "fireproof."[70] Isaiah Rogers used all these measures—interior firewalls, mortar-protected floors, and walls plastered without lath—in the Tremont House (1828–29) hotel in Boston.[71] Such steps improved a building's fire safety at a comparatively small additional cost; nevertheless, owners rarely adopted them. Such simple fire safety measures only became commonplace after cities made them part of their building laws.

Vault building construction ceased in the years after the panic of 1837, and the economic downturn in early 1840s depressed building starts. During these troubled times, vault architects subsisted on government work and the projects that were already under way. Town died in 1844, and Willard turned his attention to the granite quarries he managed. Parris devoted his final years to designing engineering works, lighthouses, and buildings at the Portsmouth Navy Yard. Walter went bankrupt and worked abroad for a period.[72]

The construction picture improved in the second half of the 1840s, during which time some of the last vaulted fireproof buildings went up. In 1846, Walter received a commission for a courthouse for Chester County in West Chester, Pennsylvania, and he designed a building with a vaulted basement and first story. This building had the standard accoutrements of fireproof buildings: iron doors hung in iron frames, iron shutters, and a metal-covered roof. Also in that year, the Boston architect and engineer George M. Dexter designed a one-story kitchen and laundry with a vaulted cellar for the Massachusetts General Hospital. Significantly, the wings he added to the hospital at the same time, which contained patient wards, were not vaulted. In 1851, Walter was appointed as the new architect of the U.S. Capitol and assigned to work on the Patent Office as well. He used the vault system to complete the east (1849–52) and west (1852–56) wings of this building.[73] Strickland's career also revived when, in 1845, he was chosen to design the Tennessee state capitol building. He moved to Nashville to superintend this large project and remained there the rest of his life. The building combined both the vault and the new iron styles of fireproof construction, containing vaulted floors and an iron roof. Completed in 1859 by Strickland's son Francis, it may be the last important masonry vaulted fireproof building constructed in the United States.

Few architects designed vault-style buildings, and those who did possessed special talents. They were technically oriented men who drew on their construction experience and ingenuity to understand, apply, and adapt construction methods that they read about or saw. While several of the early vault architects in the Philadelphia area learned directly from Latrobe, in New England, the pioneer vault builders—Parris, Town, Willard, Young—were self-taught. Nevertheless, with the exception of Robert Mills and Latrobe himself, all these men shared a background in the building trades and a willingness to implement novel structural ideas, albeit ones that had the sanction of British architectural writers. They took a chance building structures that combined massiveness and novelty, which were more likely to fail than ordinary buildings. The examples of the vault structures that survive today are monuments to their designers' skill.

With all the disadvantages, it seems hardly necessary to explain the demise of the vault system. But a final blow to the system was the change in architectural taste. As the vogue for classical revival styles faded in the 1840s—critics ridiculed the indiscriminate reproduction of Greek temples for every purpose—so did the most appropriate architectural expression for solid masonry construction. When the nation's economy perked up in the late 1840s, architects adopted a more practical system of fireproof construction borrowed from England. Most of the architects who designed vault-style buildings quickly turned to the new system, which incorporated a previously little used material, structural iron.

2

THE IRON AND BRICK
FIREPROOF BUILDING,
1840–1860

In all fireproof buildings either brick walls with wrought iron
beams, or cast or wrought iron columns with wrought iron
beams must be used in the interior.

NEW YORK CITY BUILDING LAW,
New York laws of 1860, chapter 470

As a general rule, nineteenth-century conflagrations took a heavier toll in
property than in lives. Deadly fires typically occurred in single buildings—
crowded theaters, hotels, and factories—and killed people trapped inside;
these fires rarely developed into conflagrations. In contrast, the fires that
did become urban conflagrations often started after hours and then grew
and spread, giving people time to get away. Conflagrations caused terrible
suffering nonetheless: businessmen lost their capital, workers their jobs,
and families their homes. They often caused injury far from the scene of
the fire. For example, the 1835 fire in Lower Manhattan, which destroyed
the city's warehouses and banks, disrupted trade and financial dealings of
people who conducted business in New York, wherever they lived. Nearly
all of the city's fire insurance companies shut down, some of which had
customers outside New York.[1]

This 1835 fire stands out as one of the most costly in U.S. history, but it
was just one of many conflagrations in the first half of the nineteenth century. In 1838, half of Charleston, South Carolina, burned down. Lower
Manhattan suffered a second sweeping fire, although a less destructive
one, in 1845. In the 1840s, fire leveled parts of Albany, Nantucket, Pittsburgh, St. Louis, and St. John, Newfoundland. The pattern continued in
the next decade, when Chillicothe, Ohio; St. Louis, again; Philadelphia;
and San Francisco all suffered fires.

By the middle of the 1850s, Americans began to use a new method of
fireproof construction, one that had been perfected in over a half century

of development in Great Britain. Structural iron constituted the key ingredient of the new system. Rather than cumbersome masonry walls, piers,
and columns, the new method employed iron columns, girders, and beams.
The floors between the girders and beams were brick arches, as in the
vault system, but were shallower and so took up less headroom. English
designers first introduced the iron and brick system in textile mills and
warehouses in the 1790s; by the 1840s, British architects had adopted it for
projects they might have built in the vault style. Yet American designers
continued to use the vault style for clients who wanted fireproof buildings,
although they knew about the iron and brick system. Only in the 1840s did
Americans begin to use iron structurally, by which time they could learn
from several decades of British experimentation with iron construction.

Since some technical terms will appear regularly in this section, definitions might be helpful. The words *girder, beam,* and *joist* all describe the
spanning (horizontal) elements of a structure. The word *beam* can mean
any horizontal element that crosses an opening or spans from column to
column. A beam in the shape of a capital I (hence the name *I-beam*) con-
sists of a middle section called the *web,* and perpendicular pieces at the top
and bottom, called *flanges. Girders* are large beams; usually when a large
beam supports smaller ones, the large beam is called a girder and the
smaller ones, joists. A girder that spans a long distance and is built up from
separate pieces is often called a *truss.*

Development of the Iron and Brick Fireproof
System in Great Britain

The history of the development of the iron and brick system in British industrial buildings has received much scholarly attention. The pioneer
building of this type was a cotton spinning mill in Derby, in the north
Midlands, built by the manufacturer and inventor William Strutt in
1792–93. Strutt did not invent the floor system, consisting of shallow brick
arches between beams, which he used in this mill: similar floors had been
built in Paris since at least the 1770s. Nevertheless, by using cast iron columns, rather than piers or bearing walls, to support beams, he introduced
an important new feature. This arrangement permitted a more open floor
plan, such as was difficult to achieve with the vault system, and made the
system practical for commercial and industrial buildings. Contemporaries
quickly perceived the significance of Strutt's mill: only a decade after its
completion, a contemporary guidebook called it "the first fireproof mill

that was ever built."[2] Strutt built two more iron and brick buildings in the 1790s: a warehouse in Milford and another spinning mill, the West Mill, in Belper.

What these buildings shared with vault buildings, and what qualified them to be called "fireproof," was that all structural parts were noncombustible. The prototype, the Derby mill, in fact contained timber beams, but the beams in the fireproof mills that followed were made of cast iron. An early example of an all-iron frame mill was the Ditherington mill, built in 1796–97 by Strutt's friend, the manufacturer and designer Charles Bage. Not coincidentally, this mill went up in a center of the emerging structural cast iron business. While a floor made with brick arches weighed much more than an ordinary timber-framed one, it was only about half the weight of the vaulted floors in solid masonry buildings, and therefore did not require walls as thick as those in the vault buildings[3] (Fig. 2-1). The usual features of the British fireproof mill included masonry exterior walls; cast iron beams; cast iron columns; and floors made of shallow brick arches, called segmental or jack-arches, tied with an iron bar to resist the arch's thrust. The roofs could be framed in timber or, less commonly, iron (Fig. 2-2).

For the most part, designers used iron as they did timber, slotting iron pieces together in carpentry-style joints and fastening them with bolts. A contemporary architect described the approach as "cast-iron joinery."[4] Nevertheless, iron presented a number of technical problems. For one thing, construction details for iron construction had to be prepared with great precision: ordering a piece of cast iron the wrong length would be a

Fig. 2-1. Section through a standard American type of iron and brick arch floor, made of iron I-shaped beams with a shallow brick arch spanning between them. An iron rod tied the lower webs of the beams, to resist the thrust of the arch. The top of the arch is filled with concrete to make a level surface. Wood strips, for nailing down the finish floor, are set in the concrete. J. K. Freitag, *The Fireproofing of Steel Buildings* (New York: John Wiley & Sons, 1899).

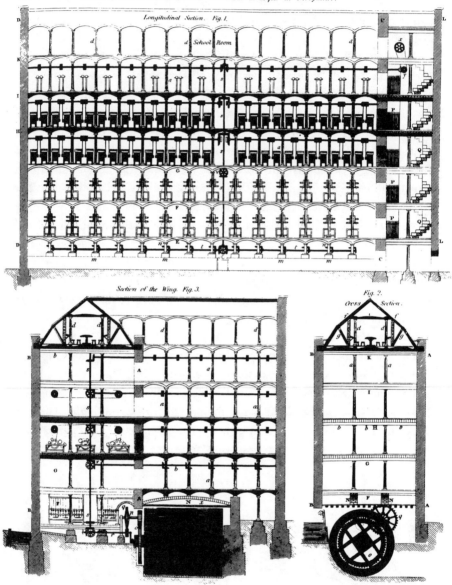

COTTON MANUFACTURE.

Sections of one of Mess.rs Strutt's COTTON MILLS at Belper in Derbyshire.

PLATE XIV.

Fig. 2-2. An English fireproof factory, in cross and longitudinal sections. Girders span the width of the mill; shallow brick arches fill between the girders. This example has an iron-framed roof and a circular stairway made of stone. The drawing is a not entirely accurate depiction of William Strutt's 1803–4 North Mill in Belper, Derbyshire. Abraham Rees, *The Cyclopaedia; or, Universal Dictionary of Arts, Sciences, and Literature* (Philadelphia: Saml. Bradford et al., 1810–24), plates vol. II, "Cotton Manufacture," plate XIV.

costly mistake. Designers had at best a rudimentary idea of the mechanics of, and safe working loads for, structural cast iron, which led them sometimes to overdesign a structure or alternatively to overload it. Finally, flaws in cast iron pieces, which could compromise their structural reliability, were difficult to detect.

Besides being tricky to design properly, an iron and brick fireproof mill cost much more than a timber-framed one, but had little to recommend it apart from fire resistance. Iron and brick did make a stiff floor, which may have been advantageous in textile mills subject to vibration from machinery. Yet iron and brick floors were heavy, took up more headroom, and, when paved with flagstones or brick, were slippery and cold. Wood floors could and did bear the working loads of mills and warehouses safely. Thus, the vast majority of British textile mills in the first half of the nineteenth century were built in the traditional way, with timber frames. Owners selected the iron and brick system when they wanted a substantial, fireproof building.

British designers introduced several variations aimed at overcoming one or another of these functional drawbacks. One variation consisted of floors made of an iron grid of closely spaced girders and joists, filled in with large flagstones, which created a thinner and therefore lighter floor. The steam-powered sawmill at the Royal Naval Dockyard in Chatham, built 1811–14, was one of the first buildings to have this kind of floor (Fig. 2-3). The system enjoyed brief popularity in Britain in the 1820s and 1830s, when it was used in some sugar refineries and small buildings, although not in any large textile mills.[5] With its closely spaced joists, determined presumably by the size of available flagstones, this system required as much or more iron than the usual iron frame, so would not have been used to save money. Americans adopted a system similar to this in the 1860s.

At the beginning of the 1820s, Thomas Tredgold, author of a popular carpenters' handbook, published perhaps the first book on building with iron, *A Practical Essay on the Strength of Cast Iron*. With its "rules, tables, and examples" for calculating the bearing strength of cast iron, it presented much-needed technical information about the structural performance of the material. Tredgold predicted a bright future for iron in buildings, because of its permanence, durability, ability to be molded, and, most of all, its "safety against fire." The book became a best-seller: a second edition came out two years after the first, followed by two more editions after the author's untimely death. It was translated into French, Italian, and

Fig. 2-3. Ceiling of a sawmill at the Royal Naval Dockyard in Chatham, England (1811–14), made of cast iron girders and cast iron joists, with large flagstones on top of the joists. From the *New Civil Engineer*, reproduced with permission from Thomas Telford Publishing, Thomas Telford Services, Ltd., Institution of Civil Engineers, London, England.

Dutch, and knocked off in a supposedly easier to understand English-language version in 1832.[6]

Architects outside the mill world began to adopt the iron and brick system. In the 1830s, Charles Barry used iron and brick floors in the Houses of Parliament, which he reconstructed after an 1834 fire, and in the Reform Club in London (1837). By the 1840s, the iron and brick system defined fireproof construction in England. An architect writing in 1872 recalled, "Among my early experiences no building was considered fire-proof that had not iron joists and brick arches."[7]

But the popularity of the system for nonindustrial buildings soon faded: by the time American architects adopted it, British architects had turned away. Exactly how many nonindustrial iron and brick buildings went up in Great Britain is unknown, but probably the number is small. Besides costing more than ordinary construction, the system gained a reputation for being dangerous. The collapse of a fireproof mill in Oldham in 1844, which killed about twenty people, received considerable publicity. Three years later, a section of a fireproof mill in Manchester collapsed.[8] These accidents stirred up apprehension out of proportion to the hazard, since they were rare events. Even the fire safety of the system came into question. One influential critic of iron and brick buildings was the superintendent of London's Fire-Brigade, James Braidwood. He argued that cast iron beams and columns, when used in large buildings without internal partitions, with open staircases and shafts, and filled with combustible goods, "are not, practically speaking, fireproof." He warned that weakened cast iron, the result of flawed casting or overloading, might go unnoticed. Fire caused iron to expand and push out walls; heated tie-rods could yield and let down the floors; and hot cast iron could fracture when doused with cold water. "For these and similar reasons," Braidwood explained, "the firemen are not permitted to go into warehouses supported by iron, when once fairly on fire."[9] If structural iron was expensive and did not create a safer building, architects reasoned, why use it?

Although it fell out of favor with British architects, mill designers stuck with the iron and brick system: in the second half of the century, it became the standard one for new factories in Britain. The men who designed iron and brick mill buildings believed in their fire resistance. Responding to criticism of the system in 1844, the prominent engineer, William Fairbairn, wrote that not only could an iron and brick warehouse be fireproof, but such buildings improved the security of the surrounding community.[10]

For Fairbairn, the sensational collapses resulted from designer error or poor management. And indeed, many British architects failed to keep up with research on the strength of and best forms for cast iron. In the early days, the two most commonly used beam sections (a section being what the beam would look like if you sliced through it vertically) were in the forms of an upside-down T or V. Functional requirements, rather than knowledge of the properties of cast iron, determined these shapes: the flanges and angled sides of the beams created a shelf for starting the arch of a brick floor.[11] Eventually, some designers realized that cast iron possessed peculiar mechanical properties and it should be used in a way that took advantage of these. Tredgold addressed this matter in his book on the strength of cast iron, but made an incorrect recommendation. Because he had little experimental data to show otherwise, he simply reasoned that cast iron was equally strong in tension as in compression. If this were so, a symmetrically shaped section was best, and he proposed that cast iron beams be made in the shape of an I.

The scientist Eaton Hodgkinson soon demonstrated the error of this conclusion. Hodgkinson had much better facilities for exploring this question: the Manchester engineering works of his friend William Fairbairn. He found that for cast iron, a T-shaped section, with a much wider bottom flange than a top one, was a stronger form than a symmetrical I-section. In other words, for a given quantity of cast iron, a beam shaped like an upside-down T could support more weight, before breaking, than one shaped like an I. Hodgkinson reported his results to the Manchester Literary and Philosophical Society, which published his paper in 1831. His work received wider exposure in a new edition of Tredgold's *Practical Essay,* published in 1842, which Hodgkinson annotated and to which he appended a section describing his experiments. A separate book about his experiments appeared in 1846.[12] His findings could be boiled down to a 7:1 ratio, bottom to the top flange, as the best shape for a cast iron beam.

American architects did not share their British colleagues' reservations about structural iron. They used the iron and brick system in nonindustrial buildings and very rarely in industrial buildings, the mirror-image British practice. And while British iron and brick factories and warehouses served as the models, Americans adapted the British systems to their own requirements. A key difference was in the floor frame. In British mills, girders typically spanned across the narrow dimension, from outside wall to columns to outside wall, spaced from about eight to ten feet apart. This arrangement resulted in rows of closely spaced columns running the length of the building. American architects built this way at first, but soon, in the 1850s, they replicated wood floor framing in iron, employing girders to carry iron joists. This form reduced the number of internal columns.

Brief Phase of the Cast Iron
Fireproof Building

Americans used iron very little structurally before mid-century, not for lack of information about how it was being used in Britain but mainly because of its high cost. Both Alexander Parris and the prominent civil engineer Loammi Baldwin Jr. owned Tredgold's textbook on iron. Parris sold his copy in 1826 to the library of the Boston Athenaeum, where it became available to the Athenaeum's members. Parris and Charles Bulfinch owned Charles Sylvester's 1819 pamphlet on the Derbyshire Infirmary, in which the author declared Strutt's method of "rendering buildings fire proof by substituting iron and brick in place of wood" to be the most important modern discovery.[13] American textile manufacturers seeking information about British production methods also learned about their methods of factory construction. For example, in 1825, before starting his career in textile manufacturing, Zachariah Allen of Rhode Island took a "practical tour" of British textile centers, which included a visit to a fireproof mill in Leeds. During his stay in Leeds, this mill caught on fire, and Allen noted that it held up well.[14]

While Americans relied on British authorities like Peter Barlow, Hodgkinson, and Tredgold for data on the strength of iron, they also conducted their own tests. The *Journal of the Franklin Institute* (JFI) reprinted Hodgkinson's seminal 1831 paper on cast iron beams a year after it came out in England, as well as subsequent articles by Hodgkinson on cast iron. In 1836, the JFI published a report on one of the earliest tests of structural iron in the United States, conducted at the West Point foundry in New York. The tests took place the year after New York City's great fire, when merchants there sought means to rebuild more securely. No one knew if American iron would perform the same as British iron. Lieutenant Thomas J. Cram, then a science instructor at the United States Military Academy, witnessed the tests and wrote up the experiment, which generally confirmed Hodgkinson's finding that in cast iron, an asymmetrical beam section was most efficient.[15]

Yet no American mill owner or city merchant actually built a British-style fireproof structure at this time, because of the high cost of iron. Once an exporter of pig iron, America fell far behind Britain in iron production in the late eighteenth century. The principal technological development that boosted British production was the introduction of smelting with coke. With their iron output growing and cost declining, the British began to apply iron to new uses, such as structural castings. Americans used coke as fuel to only a minor extent in the first half of the nineteenth century.[16] Imported iron helped fill the gap, but it was expensive, too, as a result of protective tariffs.

Still, structural iron was not entirely absent in the United States before 1850. As in England, the first form in which iron shapes appeared in buildings was as columns; these were used in churches and theaters, more because of their thin profile than for fire resistance. Since the raw crushing resistance of cast iron is considerably greater—roughly sixteen times greater—than that of even the strongest American timber, a thin iron column can do the same work as a much thicker wood post.[17] Nevertheless, some designers used cast iron columns for their fire resistance, especially in fire-prone buildings such as theaters and mills. By the late 1830s in New

York City, architects had introduced iron columns into storefronts, replacing stone posts and lintels at the ground floor with iron ones. In Boston, this new style of iron shopfront appeared around the 1840s.[18]

Bridge engineers likewise used iron very sparingly before 1850. Of the handful of iron structures from this time, probably the earliest was a small bridge built by Army engineers on the national road at Brownsville, Pennsylvania, between 1836 and 1839. A few iron bridges, also highway bridges, were built by private interests in the 1840s. In this decade, the Baltimore and Ohio Railroad began to make trestles in some of its bridges out of iron. Around 1845, the first all-iron railroad bridges appeared. In this period, engineers chose iron for permanence and fire resistance, not so much for strength.[19] Still, all-iron bridges remained the exception. As late as 1863, according to one authority, an all-iron truss bridge cost more than double what a wooden bridge of the same span and strength would cost.[20]

Iron construction in the United States took off following increases in iron output and technological advances in foundry practice, most importantly, improvements to the cupola furnace and in techniques for making large, hollow items. Pig iron production expanded in the latter 1840s, roughly doubling between 1840 and 1847. Railroads absorbed the bulk of this output, but in 1851–52, the British flooded the American market with cheap rails, which reduced the rail mills' demand for iron.[21] This released iron for other purposes, and foundries expanded to take it. Until about 1850, most foundry work occurred at blast furnace sites, where ironmakers made castings directly from iron tapped from furnaces. A cupola furnace functions like a miniature blast furnace, and was used to remelt pig iron, which both improved the iron and gave founders better control over the characteristics of the iron.[22] Better cupola furnaces allowed foundries to set up in cities, near their customers, rather than in the country, near the raw materials of iron making.

Around 1850, foundries that specialized in making structural castings first appeared. Called "architectural ironworks," they made structural elements such as columns, box lintels, girders, joists, and iron ribs for domes. They also began to make a new product: iron fronts for buildings. Builders already had introduced iron posts and lintels in ground floor shopfronts. Making iron and glass panels to cover these fronts, and then extending the panels upward to form the entire facade, seemed a logical development. Around 1849–50, such full-height cast iron facades appeared in several East Coast cities, made by foundries in Baltimore, New York, Philadelphia, and Trenton.[23] The novel fronts enclosed otherwise ordinary buildings.

The only iron parts of "cast iron architecture" were the street facades and posts on the inside; the party and rear walls were brick, and the floors and roof were timber. But because iron was noncombustible, many people assumed iron-front buildings could not burn. According to the Chicago builder and building official Henry Ericsson, when the new iron fronts appeared in Chicago in the 1850s, "people just took it for granted that such iron buildings were fireproof."[24] An architect recalled when several iron-front buildings in New York City caught fire "and the fronts fell down flat on their faces in the street, the average newspaper reporter was greatly nonplused and befogged in his graphic descriptions, and wondered how iron buildings could be burned." The answer, of course, was that the interiors of these buildings *were* combustible, and if the timber floor frame to which the iron fronts were strapped burned out, the fronts could topple. The advantages of cast iron fronts were "economy, stability, and beauty," not fire protection.[25] While architects occasionally put cast iron fronts on fireproof buildings, the typical iron-front building was no more fire-resistant than any masonry building.

Of the rare instances of iron and brick fireproof floors in the United State before the late 1840s, the earliest was a partly fireproof ropewalk (a building for the manufacture of rope) at the navy yard in Charlestown, Massachusetts, constructed between 1834 and 1837. A hazardous business, rope manufacture involved highly combustible rope stock and boiling kettles, and, at the navy yard, steam engines to drive the ropemaking machines. With its timber stores for ship building, black powder, and wooden docks with wooden ships, the navy yard could hardly afford a fire, and Navy officials ordered the engines to be enclosed in fireproof rooms. Alexander Parris, architect of buildings at the navy yard, put the boilers and engines in basement rooms at the head of the ropewalk and covered them with English-style fireproof floors, made of cast iron beams and brick arches.[26] He used cast iron beams in the shape of Is, such as Tredgold recommended, which suggests that he had not come across Hodgkinson's work. The floors over the engines were the only ones he made fireproof; the rest of the building was framed in timber (Fig. 2-4). Another building from the 1830s with iron and brick floors was the Academy of Natural Sciences in Philadelphia (1839–40). The Scottish-born John Notman, who designed the building, may have worked in William Playfair's office in Edinburgh before emigrating to the United States in 1831 and, if so, would have learned about using iron in fireproof construction there.[27] No details of the building's internal construction survive.

Fig. 2-4. Section through the basement engine room in the headhouse of the ropewalk at Charlestown Navy Yard (1834–37). The orthographic projection of the floor above the engine shows I-shaped cast iron beams and brick arches. "Plans of Buildings Erected in the Navy Yard, Boston 1830–40," National Archives, Washington, D.C.

Architects proposed using iron and brick construction for several public buildings in the 1830s and early 1840s, although the projects were not executed. For example, in his 1838 entry to the Boston customhouse design competition, Alexander Parris, having recently finished the ropewalk in Charlestown, called for making the floors of the upper two stories out of iron and brick, or else iron beams and stone slabs. But the customhouse building committee picked Ammi Young's design for a solid masonry building. In the same year, Parris recommended replacing the masonry vault floors in the Treasury building, which Robert Mills was constructing, with floors made of "cast-iron beams, with brick arches." Such floors, he wrote, were "much used in cotton-mills and other structures . . . in England" and required "no thicker wall for its support than floors of timber."[28]

Mills rejected this advice and completed the building using masonry vaults. In 1844, for a new building for the War and Navy Departments, both John Notman and William Strickland proposed iron and brick construction, but the project died. Two years later, Notman entered a plan for an iron and brick building in the design competition for the Smithsonian Institution building. He proposed to make it "thoroughly fire-proof . . . , by using cast iron beams and joists, filled in between them with brick work, constituting in fact a wall in the floor," such as he had used in the Academy of Natural Sciences.[29] James Renwick Jr., who won the Smithsonian design competition, broached the idea of introducing iron and brick floors with the Smithsonian's building committee. But, to economize, it decided against fireproof construction.

These proposals suggest that by the late 1840s, most architects considered the iron and brick construction superior to the vault system. One notable exception was Robert Mills. In discussing his design for a fireproof War and Navy building, Mills noted that other architects had proposed iron and brick construction instead of masonry vaults. He warned against using the new system, arguing that "the merits of cast iron for this purpose are greatly overrated. . . . Its use in buildings has been recently condemned in England by Parliament as dangerous to their stability."[30] Yet some in Parliament thought well enough of the iron system to allow it to be used in the reconstruction of their Houses. Mills worried that iron's expansion and contraction over time would break down a building's walls. This was a reasonable concern but one that could be addressed through design. Iron and brick construction offered so many advantages over vaults that designers were not likely to go back to the older method.

Thus, when the pace of building construction picked up in the second half of the 1840s, the iron and brick system became the new standard for fireproof construction. An 1846 article about fireproof construction in the JFI, reprinted from a British periodical, defined fireproof construction as iron and brick construction. The author's recommended building included cast iron columns; cast iron girders in a complicated shape (arched on top with a curved bottom flange); and brick arches. The iron and brick system, he asserted, was "becoming universal" for warehouses in Liverpool, where merchants had been "burned into the conviction" of the necessity of fireproof construction.[31]

The federal government lent its imprimatur to the iron and brick construction when it adopted the system for its fireproof buildings. An early example was the Winder Building in Washington, developed by a private owner for the government. Constructed between 1847 and 1848, it contained both vault and iron and brick construction. Barrel vaults supported by bearing partitions covered the central corridor on the main floor, while the rooms that lined both sides of the corridor were built in the iron and brick system. Cast iron beams spanned from the wall along the corridor to the outside wall and were filled in between with shallow brick arches tied with iron rods and turnbuckles. The floors were finished with brick and stone tile and the staircases were made of iron. The builder also installed a furnace and central heating instead of open fireplaces, a novelty at the time, for the purpose of reducing fire risk.[32]

Another early project that incorporated iron for fire safety was the second Congressional Library, built after a fire destroyed the original one on December 24, 1851. The library occupied rooms inside the Capitol and was being renovated and expanded at the time of the fire. The task of designing a new library room fell to Thomas U. Walter, newly appointed architect of the Capitol extension. Walter's design may have been influenced by the Astor Library in New York City (1849–53), then under construction, since his plan resembled it. The Astor's architect, Alexander Saeltzer, tried to make it as fireproof as possible, by avoiding wood and using iron liberally, most notably in the glass ceiling that covered the main library hall; the roof trusses; and the galleries, which contained the bookshelves.[33] Similarly, Walter made the Congressional Library ceiling of colored glass in an iron frame, supported underneath by huge iron brackets, and built iron tiers for the bookshelves and even iron shelves for the books. A skylight in the roof lit the glass ceiling from above. Janes, Beebe & Company, a New York City foundry, executed this project. Not only did the firm manufacture the iron for the library, it also installed and painted the interior of the room. They finished the work in the summer of 1853. Justly proud of his remarkable room, Walter exulted at how quickly it went up: "The whole of this immense iron room will . . . have been cast, fitted, and put up in less than *six* months; and as far as my own knowledge goes, it is the first room ever made exclusively of iron."[34]

Another early cast iron and brick arch fireproof project in Washington was the original building of the Smithsonian Institution, popularly known as the Castle. All of the entries in the 1846 Smithsonian design competition recommended some kind of fire protection for it. Robert Dale Owen, a congressman from Indiana, member of the Building Committee, and a driving force in establishing the Smithsonian Institution, had a great interest in fireproof construction. He knew about the iron and brick and the vault systems, and had read Hodgkinson.[35] But when the time came to break ground, the regents balked at the cost of fireproof construction and, thus, the building that began to go up in 1847 was of ordinary construction (masonry walls and a timber interior frame). Only a few rooms in the basement—for the furnaces, the janitor, and to store Smithson's personal effects—were vaulted. The regents authorized inexpensive fire protection measures: the attic floor was covered with two inches of cement and other floors were deafened with a layer of lime, clay, and sand. In addition, the floors in several sections were paved with stone.[36] Although well-intentioned and useful in theory, the dead weight of the mortar and stone paving increased the load on the wood structure and the exterior walls, and may have contributed to the collapse of center section of the building in February 1850. The building was still under construction at the time of the accident.

The regents then decided to reconstruct the hall fireproof. Renwick resigned as architect, and the Building Committee hired an officer in the United States Army Corps of Engineers, Lieutenant Barton S. Alexander, to design and superintend the reconstruction. Alexander's architectural experience included superintending the construction of several buildings at the Military Academy as well as the Military Asylum (Soldiers' Home) in Washington, the first rest home for Army veterans, on which he was working when the regents needed a new architect. Alexander rebuilt the center block of the Smithsonian, putting more vaults in the basement level, and iron and brick construction above. Thus, the lofty, paneled ceiling over the large room on the main level was made with cast iron beams and brick arches. He began work in 1853 and completed it in 1855. Although safety, stability, and economy dominated Alexander's design choices, he was not oblivious to architectural beauty. The decorations he chose reflected his military aesthetic: simple but "enough, as I think, to enable the building to do its duty with grace and dignity."[37] And despite his best efforts, the project ran over budget, which prompted the regents to order cost-saving changes. One was to put a wood-framed roof over the building, despite Alexander's warning that this would compromise the safety of the building.[38]

The federal government built iron and brick fireproof structures outside Washington as well. One of the most remote of these was the customhouse in San Francisco. The Treasury Department wrestled with how to get a secure structure for itself in this distant and fire-prone boomtown. After dropping the idea of shipping out an all-iron building, because of the high cost, the Department then asked Thomas Walter to make a design. In 1851, he planned a partly fireproof building; it included a vaulted cellar and collector's room, and elsewhere, wood floors protected with three inches of concrete between the joists. The attic floor was to be iron and brick, "so as to form an entire brick covering over the whole building under the roof."[39] A Boston firm won the contract to supply the iron for the building, which was to be made under the supervision of architect Gridley J. F. Bryant. Bryant had worked with Alexander Parris at the Charlestown Navy Yard in the 1830s, when Parris built the ropewalk with its iron and brick arched floors. After learning about two sweeping fires in San Francisco in 1851, Bryant urged the Department to make all the floors iron and brick, and the Department agreed.[40] Completed in 1855, the customhouse was being demolished in 1906, to make way for a larger structure, when the earthquake and fire struck the city.

State and local government adopted the iron and brick system for their buildings. An early example was the fireproof statehouse in Columbia, South Carolina. A circa 1848 contract for cast iron girders for this building went to Hayward, Bartlett & Company, a Baltimore foundry that became one of the nation's leading architectural iron firms. In Massachusetts, three buildings—an addition to the statehouse, the public library in Boston, and the city hall in Springfield—all contained iron and brick floors. Gridley J. F. Bryant, who was supervising the iron for the San Francisco customhouse, designed and built the statehouse wings (1853–54), and made the third and attic floors out of iron and brick. In Springfield's partly fireproof city hall, designed in 1853, the New York architect Leopold Eidlitz used iron girders in the first floor and in the roof frame.[41]

Businessmen, who for the most part never used vault construction, adopted the iron and brick system for their buildings. An early fireproof commercial building was the plant of I. M. Singer & Company, manufacturer of sewing machines, on Mott Street in New York City. Its architect, George H. Johnson of the Architectural Iron Works in New York City, made the floors on all six stories, as well as those over two basement levels and in an engine house at the back, of iron beams and brick arches. Architectural Iron Works, which supplied the iron for the building, described it in an 1865 advertising catalogue as "perfectly fire-proof; the roof, the front and rear, including sashes, shutters, &c., being entirely of iron. The floors are supported upon iron columns and girders, with iron beams and brick arches. The stairs, &c., are of iron."[42] In Cincinnati, Isaiah Rogers designed iron and brick floors for his Commercial Bank (built in 1849).[43]

This cast iron phase of iron and brick construction was a transitional one in which no single form of construction became standard. Architects combined iron and brick with vaults and ordinary construction, and experimented with various shapes for iron beams and fillings between beams. The entries in an 1855 competition for a new "completely and absolutely fireproof" library for Boston illustrate the variety of iron floors in this period. In one entry, Edward and J. C. Cabot proposed floors with stone slabs rather than brick arches between cast iron beams. John R. Hall's entry called for floors made of cast iron joists and iron plates. A drawing by the winning architect, Charles Kirby, dated after the building was completed, shows symmetrical, I-shaped cast iron beams and brick arches at the ground level and flat floors on the upper levels.[44] The building was actually constructed with girders shaped like upside-down Ts, sixteen inches deep and spaced about ten feet apart. On the flanges of these girders were smaller beams shaped like inverted Ys, roughly nine inches deep and spaced about four feet apart, filled between with brick arches[45] (Fig. 2-5).

In the mid-1850s, a standard system of iron framing for fireproof buildings emerged, one that incorporated wrought iron in place of cast iron joists. The federal government became an important disseminator of the system when, at this time, it began to put up new buildings at an unprecedented rate. Except for a few small buildings on the frontier, the government ordered the new public buildings to be fireproof, and the Treasury Department adopted the iron and brick system to build them.

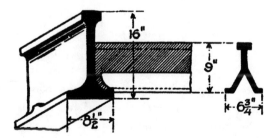

Fig. 2-5. Cast iron floor members from 1855, used in the Boston Public Library building: girder (*left*) and Y-shaped "vault beam" joist (*right*). Joseph K. Freitag, *Architectural Engineering* (New York: John Wiley & Sons, 1904).

The Federal Government Adopts Iron
and Brick Fireproof Construction

The government's huge construction program in the 1850s resulted from several factors: the emergence of new cities; the need to replace outdated buildings in the growing cities in the East; and, most important of all, a flush Treasury. Federal receipts jumped by more than a third between 1849 and 1850 and increased another 20 percent in 1851, principally from customs revenue. Congressmen developed a taste for pork and began to authorize new buildings at a great rate. By 1853, seventeen customhouses and five marine hospitals were under construction, in addition to an Assay Office in New York City. In its first session in 1854, Congress voted for sixteen more customhouses and six marine hospitals to be built in as many different cities, from Belfast, Maine, to Galveston, Texas, to Burlington, Iowa. In all, Congress approved about seventy buildings for civil purposes in the 1850s. The responsibility for erecting these fell to the Treasury Department.[46]

With the workload escalating, Treasury Secretary Thomas Corwin decided to put an architect on staff. Before 1852, private architects typically designed federal buildings and ad hoc building committees, made up of the local customs collector and politically connected businessmen, supervised the construction. The committees were independent and could not be held accountable, for example, for delays or cost overruns. Corwin tried to improve the process by putting the architectural work, at least, in the hands of an employee. He appointed Ammi B. Young, architect of Boston's customhouse and other public buildings, for the newly created position, supervising architect of the Treasury.[47]

Although he was an expert in vault construction, Young promptly adopted the new iron and brick system for the upper floors of the federal buildings Corwin assigned him. One of his early projects was the Cincinnati customhouse (1852–57), designed by local architects and already under construction. Young revised the plan of this classical revival-style building, inserting an iron and brick fireproof frame behind the temple walls. He made the frame of cast iron columns, girders, and beams, and the floors of brick arches leveled up with "cement mortar, brickbats, coarse gravel, &c."[48] The building contained a considerable amount of iron in addition to that in the frame: the stairways; cornice and balustrade; window and door frames; capitals and bases of wall columns; balcony; angle-sided blocks for starting the arches (called skewbacks); door thresholds;

and window shutters were made of iron. These novel materials behind the walls had no impact on the facade. Young's specifications and drawings show that he was acquainted with English fireproof construction practice. His 1855 drawing for the building shows asymmetrical cast iron beams—larger flange on the bottom than on the top—such as Hodgkinson recommended. Young's specifications also required that girders and beams be "tried and proved" before going into the building, a common provision in British specifications.[49]

While iron remained an expensive material, the iron and brick system was cheaper than vaulted construction. The cost difference is indicated by the bids for the Cincinnati customhouse: one contractor proposed to build it in the vaulted style for $169,000 or in the iron and brick style for $138,000. The actual cost of the structure came to about $132,000, with stone representing 31 percent of the total cost, and the iron, about 29 percent.[50] Other buildings designed or started during this cast iron period (roughly 1852–53) include customhouses in Bath, Maine; Mobile, Alabama; Norfolk, Virginia; Pittsburgh, Pennsylvania; and Wilmington, Delaware. Young revised his 1850 design for the customhouse in Charleston, which had been vaulted throughout, to make the upper floors iron and brick.[51]

The fact that few tradesmen had any experience with this type of construction does not seem to have delayed these projects. In Cincinnati, for example, only one firm bid on the contract to make iron for the customhouse. Likewise, one firm bid on the tin roofing contract and one bid to supply bricks. The government hired these sole bidders—all from Cincinnati—and the project went forward. Horton & Macy, the foundry that made the iron, supplied fifty-two iron girders and sixty-six beams ranging in depth from one to one-and-a-half feet; this was a remarkable achievement since the firm could not have had much experience making large, structural castings.[52] Equally impressive, the workmen employed on the building do not seem to have had any special problems putting it up.

With a growing number of projects in the works, James Guthrie, who took over as Treasury Secretary in 1853, decided to centralize additional aspects of the development process. He created an office variously called the Construction Branch, Bureau of Construction, or the Office of Construction to administer and support this work. Along with design, which the supervising architect already handled, the new office had responsibility for site selection; bidding and contractor selection; and payment approvals and accounting. To head the new office, Guthrie sought a "scientific and practical engineer" and applied to the Secretary of War for such a man.[53]

Although trained to construct supply routes and coastal fortifications, military engineers became more involved in civil projects following the General Survey Act of 1824. The Act allowed Army engineers to make surveys for and direct the construction of private projects, such as roads, canals, navigation improvements, and railroads. After its repeal in 1838, engineer officers continued to work on civil construction projects for the federal government, such as lighthouses and public buildings. They had a reputation for being insufficiently cost-conscious, but were considered trustworthy, "above suspicion in money matters," which was a vital quality for construction managers to have.[54] In 1853, several Army engineers already were engaged on federal building projects. These included Captain Alexander, then constructing the Smithsonian in Washington; Captain Montgomery C. Meigs, recently given responsibility for contracts and disbursements for the extension of the U.S. Capitol and Washington post office; Major P. G. T. Beauregard, construction superintendent for the third New Orleans customhouse; and Captain Danville Leadbetter, superintendent of the Mobile customhouse. The characteristics Secretary Guthrie sought in an engineer to manage the Treasury's building program—"scientific and practical"—were usually considered antithetical. Nevertheless, in Captain Alexander Hamilton Bowman, Guthrie found his ideal man. Appointed to the Engineers Corps after graduating third in his West Point class of 1825, Bowman worked at a variety of engineering tasks, including river improvements, roads, and fortifications, principally in South Carolina.[55] The Secretary of War assigned Bowman to the post engineer-in-charge in June 1853, and he took up his duties a few months later.

Ammi Young, whom Guthrie kept on as supervising architect, now reported to Bowman, and the two men divided the responsibilities of the office. Bowman managed the development process overall, from evaluating potential building sites to communicating with Congress about construction progress and finances. Young's duties were "to make all the designs, plans, and working drawings of all the Buildings being erected under the direction of the Secretary of the Treasury," along with construction specifications.[56] The office increased its staff of draftsmen and clerks, and hired superintendents for the job sites.

One reform Guthrie sought was uniformity in the construction process, and to accomplish this, Bowman established policies and procedures that are much like those used on government contracts today. He required public bidding and sealed bids; written contracts; performance bonds; regular progress reports from project superintendents; regular financial reports from the disbursing agents; and verifications that the work performed corresponded to specifications and prices in the contracts. He outlined the procedures in his "regulations for the construction of Custom Houses and other buildings," which he sent to all construction superintendents.[57] Extant records show that during Bowman's tenure, superintendents complied with his reporting requirements.

The rationale for centralizing various aspects of the construction process was to better manage expenditures and schedules, but this reorganization also had important consequences for building design. All designs and specifications were issued by one office, under the direction of one architect and one engineer, for a limited range of building types, all of which were being built almost at the same time. The result was that uniform designs and specifications evolved, one for each kind of building: for example, there was a basic design for customhouse-post office-federal courthouse structures and another for marine hospitals. Young enlarged or shrank the model depending on the amount of interior space required at a particular location. Facade ornamentation and materials varied somewhat from place to place. On the inside, however, the materials and structural systems were highly uniform. In 1854, the Department's construction specifications called for wrought iron beams in place of the cast iron ones.

While nearly all of the iron used structurally before the 1850s was cast iron, builders also employed another form of iron, called wrought iron, to a limited extent, for such things as clamps, tie-rods, and nails.[58] Cast iron and wrought iron differ in their mechanical properties. The small amount of carbon found in cast iron, usually 5 percent or less, lowers its melting temperature, which allows the iron to be liquefied and poured into molds. Carbon makes iron very strong in compression, but it also makes it brittle and inflexible. Removing the carbon creates wrought iron, which, as its name suggests, can be bent and worked. Wrought iron is shaped by heating it until it softens, then hammering, squeezing, and rolling it. Its malleability gives it good tensile strength—allowing it to bend without breaking—but it lacks the compressive strength of cast iron. Techniques for making large structural shapes with cast iron emerged much earlier than those for rolling large shapes in wrought iron, which is why the earliest beams and columns were made of cast iron. In addition, because of the extra steps required to produce wrought iron—to remove the carbon and other impurities from the furnace iron—it cost more than cast iron.

At first, builders did not appreciate the distinct mechanical properties of wrought and cast iron. Although French designers used wrought iron in trusses in the late eighteenth century, they may have done so because they considered wrought iron a high-quality material, not because of its tensile strength.[59] With the coming of railroads and iron rails, railroad men noticed that wrought iron rails held up much better than cast iron ones. Then in the 1840s, while planning two railroad bridges in Wales—the Britannia and the Conway—the British engineers Robert Stephenson and William Fairbairn demonstrated experimentally that wrought iron withstood bending pressure much better than cast iron, although compressive force less well.[60] They concluded, moreover, that the best section for wrought iron spanning members was a symmetrical one, with reinforcement at the top to prevent crushing or buckling. The forms of wrought iron girders they recommended were the "box beam" ("rectangular tube") and "plate girder," the latter in the shape of an I (Fig. 2-6).

Fairbairn collected the growing body of information on wrought iron and iron construction in his 1854 treatise, *On the Application of Cast and Wrought Iron to Building Purposes*. The "building purposes" Fairbairn had in mind was the construction of fireproof buildings, these being the

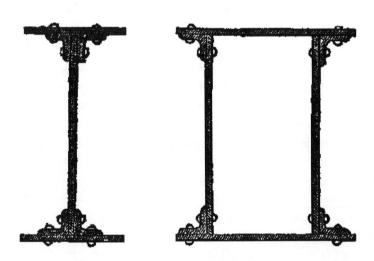

Fig. 2-6. Early structurally reliable forms of wrought iron girders: plate beam or girder (*left*) and box girder (*right*). William Fairbairn, *On the Application of Cast and Wrought Iron to Building Purposes* (New York: John Wiley, 1854).

only kind that incorporated iron girders throughout. Fairbairn urged builders to adopt wrought iron beams: they were "safer" (less susceptible to manufacturing imperfections), stronger in tension for a given weight, and could be made in lengths that permitted much longer clear spans. For these reasons, he recommended them "in substitution of the more cumbersome and uncertain ones of cast iron now in general use." Yet wrought iron girders cost more than cast iron girders of comparable bearing strength: wrought iron plate and box girders had to be fabricated by riveting pieces of iron together, a labor-intensive—hence costly—process. A solution to this, for small beams, was to roll them in one piece, which Fairbairn expected British ironmakers would eventually be able to do.[61] Apparently Fairbairn did not know that solid I-beams were already available and being used in Britain in 1854. Ironmakers in France began rolling solid I-beams in the latter half of the 1840s. Exactly when British mills first produced them is uncertain. Whether domestic or imported, British designers used I-beams in the early 1850s, for example, in a kind of iron and concrete fireproof floor called the Fox and Barrett floor.[62]

Around this time, two American ironmakers, Trenton Iron Works in Trenton, New Jersey, and Phoenix Iron Company, in Phoenixville, Pennsylvania, began to manufacture solid rolled beams. American mills had only recently managed to master iron rolling. Until the mid-1840s, all the wrought iron rails used in the United States had been imported from Britain. By the end of the 1840s, about fifteen American mills were rolling rails, although railroad managers considered American-made rails inferior to British rails and continued to import rails. In the early 1850s, British manufacturers dumped excess rails in the United States, following a slowdown in railroad growth at home. Beams, in contrast, made by the same rolling process, faced little foreign competition, while the demand for them by owners wanting fireproof buildings could only increase.[63]

Thus, just when the federal government's need for structural iron was exploding, American-made, solid wrought iron beams became available. The government was one the earliest and, in this decade, largest customer for rolled iron beams. The Treasury Department promptly substituted rolled iron for cast iron beams. After they became available, buildings under construction that had been designed to have cast iron beams were completed with rolled beams.[64]

The process of producing solid beams had not been easy for the Trenton Iron Works (TIW), the first mill, to do it. TIW was owned by the firm Cooper & Hewitt (after 1857, Cooper, Hewitt & Company), whose active principals were Edward Cooper and Abram Hewitt, son and son-in-law, respectively, of the inventor, philanthropist, and the company's largest stockholder, Peter Cooper. Cooper & Hewitt had been trying to roll solid wrought iron beams for fireproof buildings since 1852, but as of autumn 1853 and despite an investment of $150,000 in the effort, they had not achieved their goal. The best they could offer customers was "compound" beams made of two U-shaped bars, called channels, bolted together in the middle[65] (Fig. 2-7).

These were the sort of beams Cooper & Hewitt proposed to Captain Bowman in 1853 for the fireproof Assay Office in New York City, his first project as engineer-in-charge. The Assay Office, built next to the customhouse on Wall Street, was essentially a small foundry, where the gold flooding in from California was refined and stored. Initially, Bowman planned to make the building entirely of iron, but opted for brick walls when the iron bids exceeded his budget. Cost may also have played a part in his decision to substitute wrought iron for cast iron beams. The plans called for huge cast iron, bowstring girders, measuring more than two feet deep and thirty-five feet long, spanning the full width of the structure. Before he ordered these girders, Bowman learned about Cooper & Hewitt's wrought iron beams, which cost about half as much as the cast iron gird-

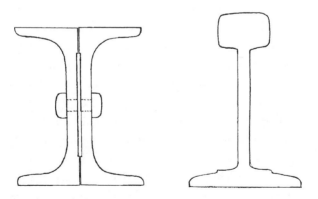

Fig. 2-7. Cooper & Hewitt's early wrought iron beams. Compound beam, first produced in 1853, made of two channels bolted together (*left*). The earliest form of a solid rolled beam, shaped like a railroad rail (*right*). Drawn by Cooper & Hewitt, August 1854. Office of the Architect of the Capitol, Cooper & Hewitt letter to M. C. Meigs, August 12, 1854, Capitol Extension—Iron Beams-Girders, 1855–71, box 9.

ers. He tested one for strength by suspending a nine-thousand-pound weight at the middle of its fifteen-foot length and found it suffered no permanent deflection.[66] Impressed, he decided to use these beams; but since TIW could not produce them in thirty-five-foot lengths, Bowman redesigned the interior and placed a row of columns along the center of each floor to accommodate shorter beams.

By early 1854, Cooper & Hewitt finally succeeded in rolling solid beams. The beams were in the shape of a railroad rail, with a rounded top and flanged bottom.[67] While this represented a great achievement, they were not in the symmetrical I-section, which was what the mill intended to make. Customers nevertheless preferred the solid beams to the double-channels. In the summer of 1854, Cooper & Hewitt claimed they could produce these rail-shaped beams at the rate of fifty tons per day.[68]

Despite their rail-like shape, these seven-inch beams were much taller than rails and, at twenty-eight to thirty pounds a foot, heavier, too. Yet their small size limited their utility. Thus, while Cooper & Hewitt worked to roll beams in a true I-shape, they also tried to make them deeper. In October 1855, a year and a half after shipping the first rail beams, Trenton announced it could roll eight-inch deep beams, and by March 1856, nine-inch beams, although still in the rail pattern. Finally, in the summer of 1856, the mill produced I-section beams, nine inches deep. In announcing the availability of these new beams to Captain Meigs, Cooper & Hewitt made no secret of their frustration: "After delays intolerable and losses too great to be borne, we have succeeded in making 9 inch beams."[69] The I-beam quickly replaced rail-beams in federal building construction. Nevertheless, mills continued to roll T-shaped beams for shipbuilders, and they came to be known as "deck beams."

At about this time—in late 1856 or early 1857—Phoenix Iron Company also began to produce solid beams. Phoenix Iron Company was one of the larger ironmakers in this period and had been rolling railroad rails for about ten years. Exactly when the first beams came off of Phoenix's rolls is uncertain, but by August 1857, the company claimed to have rolled "several thousand beams."[70] Phoenix experienced less difficulty than Trenton did getting its beam mill functioning successfully, and it also could make large beams at an earlier date than Trenton. Much of the credit for Phoenix's success goes to John Griffin (also spelled Griffen), its production superintendent, who introduced important innovations. The common method for making long rails, as well as beams, was to weld together a stack of flat bars (called a pile) and squeeze them through the rolls. This process strained the metal at the top and bottom, in shaping the flanges. Griffin introduced the idea of making pieces in the rough shape of flanges and placing these on the top and bottom of a pile (which formed the web), then putting the stack through the rolls.[71] This resulted in fewer ragged-edged and defective beams, and was the invention that gave Phoenix the technological edge over TIW. In the summer of 1857, as TIW tried to roll beams deeper than nine inches, Phoenix announced that it planned to produce eighteen to twenty-four-inch-deep beams and to make truss beams on the Zores (French) pattern.

Large beams could serve as girders, in place of the fabricated wrought iron or cast iron girders then in use. Cast iron girders were large and heavy compared with wrought iron; but fabricated wrought iron girders were expensive, and the ingredients for making them—angles and channels— were in short supply. Designers improvised economical wrought iron girders. An engineer writing in 1903 remarked on the "curious forms" of them that dated from this experimental period: "The idea of building up I-shaped girders from the wrought iron shapes then procurable led to examples so strange and so dependent upon faith that the wonder is that they even continued to stand"[72] (Fig. 2-8). Because of the cost and difficulty of making girders in wrought iron, designers continued to use cast iron, although they had largely abandoned cast iron for joists. Even after rolling mills could produce solid beams large enough to serve as girders, they tended to make them only for large orders, so small projects still had only fabricated wrought iron or cast iron girders to choose from.[73]

The Assay Office served as a training ground for builders in the methods of iron and brick fireproof construction. All of the iron contractors who worked on this project went on to become big names in New York City's architectural iron industry. James Bogardus supplied the iron cornice, and perhaps also the window sashes and frames; George R. Jackson & Company supplied the iron stairs and shutters; S. B. Althause & Company made the iron floor; and Marshall Lefferts and Brother made the iron roof. Bowman took pride in his creation: "wrought iron beams, with segmental brick arches, were used for all the floors. The shutters, doors, sash and stairs are iron so that the building is perfectly fire-proof."[74] In late 1853, Secretary Guthrie ordered henceforth, the Treasury Department's fireproof buildings would have "Brick walls, Rolled Iron Beams, Iron Roof & c."[75] (Fig. 2-9).

The iron roofs of the federal buildings were as novel as their iron and brick floors. Noncombustible roof coverings, such as slate and tile, pro-

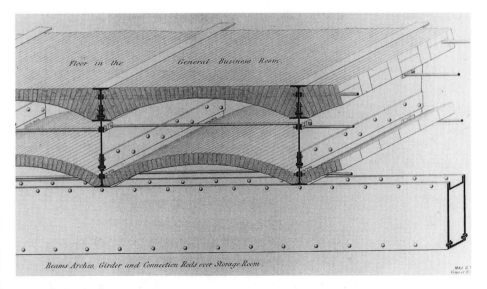

Floor in the *General Business Room.*

Beams Arches, Girder and Connection Rods over Storage Room.

Fig. 2-8. Improvised girders supporting a brick ceiling and above it, a brick floor, designed for the New Orleans customhouse, circa 1857. The fabricated girders consisted of two T-beams, top and bottom, spliced to a web plate. The splice on the top was made with angle irons, which served as bearers for the floor arch. A box girder supports the whole assembly. New Orleans customhouse, "Plans of Public Buildings in Course of Construction for the United States of America Under the Direction of the Secretary of the Treasury," 1855–56.

vided some protection against external fires. In the federal buildings, the frame as well as the covering was noncombustible. Iron-framed roofs were first introduced in France at the end of the eighteenth century, when architects put iron roofs on a number of buildings, notably fire-prone theaters. British mill owners placed iron roofs on fireproof mills in the first decade of the nineteenth century; in the 1820s and 1830s, they appeared on military and gas manufacturing buildings in Britain. Perhaps the first iron roof in the United States was one erected around 1835 by Samuel V. Merrick for a new gas works for Philadelphia. Merrick, who became a prominent iron founder in Philadelphia, visited England, France, and Belgium in 1834 to study gas installations and would have seen the European examples. In the 1840s, architects Alexander Parris and John Notman recommended iron roofs for the fireproof buildings they designed; for authority for this idea, Parris cited the iron roofs on public buildings in England.[76]

The style of iron roof adopted by Treasury Department in the 1850s consisted of a frame made of cast and wrought iron and a covering of galvanized, corrugated iron sheets riveted at the edges. The corrugations stiffened the sheets, making them self-supporting. Marshall Lefferts & Brother, an architectural iron foundry and early importer and manufacturer of galvanized iron, manufactured the Assay Office roof, and featured it in the firm's 1854 trade catalogue. All of New York's leading architectural iron companies—Bogardus & Hoppin, G. R. Jackson & Company, and Janes, Beebe, in addition to Lefferts—made proposals for the Assay Office roof, so presumably all were ready and able to make iron roofs. Around this time, Hayward, Bartlett & Company began to manufacture a roof similar to the

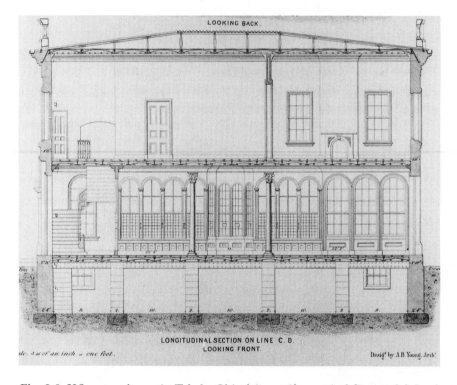

Fig. 2-9. U.S. customhouse in Toledo, Ohio (1855–58), a typical fireproof federal building of the 1850s. Longitudinal section showing the arch brick floors and iron roof frame covered with corrugated iron sheets. In the drawing, the floor beams are shown as the rail type. Toledo, Ohio customhouse, "Plans of Public Buildings in Course of Construction for the United States of America Under the Direction of the Secretary of the Treasury," 1855–56.

Lefferts roof.[77] Later, both the Trenton and the Phoenix mills manufactured iron roofs.

Through its large and steady demand for structural iron, the Treasury Department encouraged the development of the structural metal industry. Congress directed the administration to buy only from American mills and kept watch to see that it complied. A well-organized pressure group, American ironmakers lobbied relentlessly for high tariffs on iron imports. Bowman supported the buy-American policy; he assured the iron founder, Marshall Lefferts, that the Treasury Secretary "is aiming by all legitimate means in his power to favor this very important branch of American Industry."[78] The Department ordered all beams required for every project directly, thereby obviating the possibility that contractors might use imported beams. Bowman declared in his annual report for fiscal year 1858 that "the enormous consumption of iron by the government is materially promoting the . . . welfare of the great interests involved in its production."[79] Other federal projects not under control of the Treasury—the extension of the U.S. Capitol, Washington post office, and Patent Office— also consumed vast quantities of American iron beams. This steady patronage helped Trenton Iron Company establish its beam business and later weather the panic of 1857, when orders from private customers declined.[80] Phoenix Iron Company, which entered the beam business a little later than Trenton and whose principals were not politically connected to the party in power, as Cooper and Hewitt were, received less government patronage. Phoenix tried to get federal work and believed this business would help it develop its rolling mill technology.[81] As Samuel Reeves, the firm's vice president, wrote to Captain Meigs, "If the government will give us an order for solid girders to an extent to justify us in preparing the work, we will engage to produce [them] of a depth of 12 inches and 15 inches of the proper sections 35 feet long."[82] Reeves did manage to obtain beam contracts from Captain Meigs, who was not politically allied with the current administration, for the projects Meigs supervised.

Another way in which the government helped the structural iron industry was by conducting tests of iron and disseminating the results. Designers lacked information about the strength and durability of rolled iron beams. Bowman acknowledged that he had probably used too many beams in the Assay Office. Overdesign like this wasted both materials and money. In 1854, with many fireproof buildings in the works, Bowman requested funds for experiments on iron that would help the Department use iron efficiently. He noted that private interests would also benefit from the results: "As a matter of general interest to builders, [these tests] could not fail to be a popular measure."[83]

Probably the first load tests of American wrought iron beams were those Cooper & Hewitt conducted for Bowman in 1853 on their compound beam. Although very rudimentary—they simply suspended various weights from the middle of a beam—the test nevertheless convinced Bowman that it was sufficiently strong for the Assay Office. Cooper & Hewitt conducted a similar sort of test on their new seven-inch high, rail-headed beam in 1854.[84] A more comprehensive series of experiments, on solid beams and box girders, took place at Trenton Iron Works from October to the end of December 1854.[85] When other manufacturers offered the government rolled beams, samples were sent for testing, usually to Trenton Iron Works, where an agent of the Treasury Department, Major Robert Anderson, was stationed for the purpose of inspecting iron for government buildings. Thus, in 1857, Phoenix Iron Works sent beams to Trenton for testing; Bowman invited Morris Jones & Company to send its girders (fabricated ones, not solid beams) to Trenton to be tested.[86] In 1859, Major Anderson conducted experiments to determine the elasticity of rolled iron beams. Captain Meigs also tested the iron of various manufacturers, although only specimens rather than full-sized members.[87] The results of these tests became known to building interests through professional papers and the trade circulars and handbooks issued by rolling mills. In an example of the former, architect Robert G. Hatfield prepared a paper on how to calculate the required distance between centers for beams to support a given floor load, in which he used data from the 1859 Trenton tests and from similar tests made on Phoenix Iron Company beams.[88] The American Institute of Architects published and distributed this 1868 paper.

In Bowman's view, by helping to develop fireproof construction, the government rendered an important service to the nation and he endeavored to publicize the projects. The Construction Office freely distributed bid documents, which included building plans, to prospective contractors; they could learn about the details of fireproof construction from these fine lithographs. To reach a wider audience, Bowman collected sets of the building plans and sent them to "principal colleges in each state," to public libraries around the United States, and to representatives of foreign countries. Also intending to show American progress in the building arts, Bowman transmitted the volumes with a letter that explained, "The introduction of wrought iron beams and girders it is believed is new, and so far has been entirely successful."[89] The Construction Office served as a clearinghouse for information on new materials: it answered inquiries about new building materials and methods, received information on new inventions, and maintained a sample room with materials from all over the United States, which the public could visit. Bowman considered the sample room the best place for an inventor to make his products known.[90] But the demonstration effect of the actual buildings served as the most important means of diffusion. The Department erected iron and brick fireproof buildings in cities all over the country, completing nearly sixty between 1855 and 1860.[91] Not only could local builders see them go up, but many also worked on them and thereby gained experience they could transfer to private projects.

Increase in Fireproof Construction

Following the introduction of the wrought iron beam, private owners began to put up iron and brick fireproof buildings. When built with rolled iron joists, iron frames more nearly resembled timber frames than cast iron ones did. Ironmakers and the technically minded architects who designed fireproof buildings promoted the system, persuading clients to adopt it. By March 1855, Cooper & Hewitt claimed it had "orders on hand for over fifty buildings, although we have been able to make beams only about 6 mos." Many of these were federal buildings, but a notable share was private. Practically all of Trenton's rolled beams went into fireproof buildings.[92]

The first two private buildings to use Cooper & Hewitt's wrought iron beams were the Cooper Institute and the new offices and shops of Harper & Brothers, publishers, both in New York City. Construction on these buildings started before Trenton Iron Works could produce solid rolled beams, and therefore compound beams were used in the lower floors; the upper stories of both were finished with solid beams.[93] Although under construction at the same time, the Harper building came to be better known as a model fireproof building, described by a contemporary architect as "one of the pioneer buildings of the new dispensation."[94] After fire destroyed their original plant in December 1853, the Harper brothers resolved to make their new one fireproof.[95] John B. Corlies, the architect and builder they hired to erect the new building, might have put up one with a cast iron frame, but before he could do so, Abram Hewitt convinced the Harpers to use his company's new rolled iron beams.

An 1855 book about the Harper firm devoted many pages to the firm's new home and thereby preserved construction details of this important building, demolished long ago. In fact, Hewitt wrote the chapter titled "The Fire-proof Floors," in which he contrasted the ponderous vault style unfavorably with the iron and brick system. The floors in the Harper building consisted of girders made of cast iron, arched and tied across the bottom, with brackets spaced every four feet along the top that held the wrought iron beams (Fig. 2-10). Corlies treated these large girders as architectural features; he designed highly ornamental ones for the public rooms and plain ones for the workshop floors (Fig. 2-11). In addition to using only noncombustible materials, Corlies included another novel measure to safeguard the plant. He put up two buildings, one fronting Franklin Square, the other on Cliff Street, and separated them with a courtyard; in this courtyard he built a tower with a stairway, accessed via iron bridges at each floor. Only this stairway ran the full height of the buildings. By removing stairways from inside the buildings, Corlies eliminated them as conduits for spreading fire. Yet this single stair-tower served two crowded buildings and would surely have been inadequate as an exit in an emergency.[96]

These two buildings were the start of a spurt of development of fireproof buildings in New York City, making it the center of fireproof construction. A virtuous cycle took hold in which materials manufacturers, architects, and tradesmen spread information about the new technology, gained experience on fireproof projects, and then used this knowledge in new projects. Fireproof buildings from this period include the Mechanics Bank, attributed to Richard Upjohn, and the Continental Bank Building and the American Exchange Bank Building, both designed by Leopold Eidlitz. Both Upjohn and Eidlitz worked on iron and brick buildings before the arrival of rolled beams. When the architectural writer Montgomery Schuyler called the American Exchange Bank the "first fireproof building erected for commercial purposes in New York," he presumably meant that it was the first fireproof *office* building, since the Harper & Brothers' building predated it by a year.[97] The owners of the *New York Times* made their new building at Park Row "partially fire-proof" by installing an iron and brick floor over the basement level where the newspaper's printing operations were housed. This floor protected the tenants on the upper stories from a fire in the pressroom. A contemporary architectural writer considered the Metropolitan Gas Works (c. 1861–62) to be one of the best of the several fireproof gas works in New York.[98] The Astor Library added a fireproof extension, for which it ordered ninety nine-inch beams from

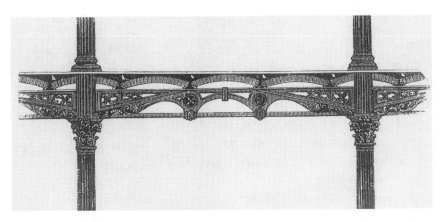

Fig. 2-10. Bowstring, or "tension rod," girder designed for the Harper & Brothers' building, New York City. Brackets for carrying the rolled iron beams (shown as the rail type) were cast on the top, and brick arches span between these beams. Jacob Abbott, *The Harper Establishment; or, How the Story Books Are Made* (1855).

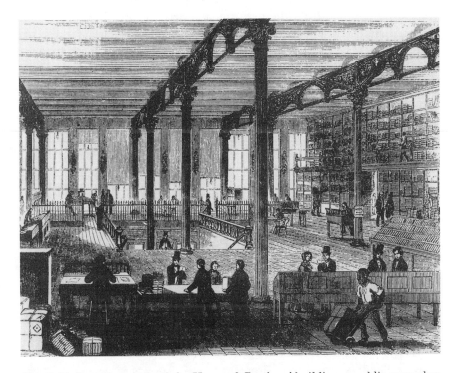

Fig. 2-11. Counting room of the Harper & Brothers' building, a public room that overlooked Franklin Square. The building's iron framework was proudly displayed, since no one in the 1850s realized that exposed iron was vulnerable to damage in a fire. Jacob Abbott, *The Harper Establishment; or, How the Story Books Are Made* (1855).

TIW in 1857. Tatham & Brothers's lead works likewise bought beams from TIW for its new, and presumably fireproof, lead warehouse.[99]

A few privately owned iron and brick fireproof buildings appeared outside New York as well. While it might be expected that Philadelphia would have a fair number, considering the tradition of vault construction there and its proximity to both Trenton and Phoenix Iron Company's mills, only a handful of fireproof buildings went up there in the 1850s. The earliest of these was a new factory (1855–56) for the prominent lamp manufacturers, Cornelius & Baker. Like the Harper brothers, Cornelius & Baker built their fireproof factory after fire destroyed an earlier plant, in December 1854. In praising the new building, a contemporary writer exclaimed, "not a pound of nails nor a particle of wood was used in the construction of the building, it is all composed of brick and iron . . . ; the floors are of brick, the stairs and window sash of iron, and the roof of slate and iron." He went on to call it "the first fire-proof building erected in the city," by which he meant the first iron and brick one.[100] This factory had the distinction of being one of the few fireproof industrial buildings in the United States. Investors put up two iron and brick banks: the Bank of Pennsylvania (1857–59), later the Philadelphia Bank, and the Consolidated Bank.[101] The two earliest private buildings to be constructed with beams from the Phoenix Iron Company (rather than TIW) were in Philadelphia: the fireproof Kearsley Home (1861) at Christ Church Hospital, and an office building (c. 1857) for the Pennsylvania Railroad, later known as the Lehigh Valley Railroad building.[102]

Another project from this period that used Phoenix's beams was a hospital for the mentally ill in Baltimore, Shepard's Asylum, later named the Shepard and Enoch Pratt Hospital. The trustees wanted it to be made "fireproof as far as possible." Thus, the first section of the main building, the Western Division, constructed in the 1860s, has iron and brick floors.[103]

New Englanders, never enthusiastic about vault buildings, put up few examples of the new type. These included the Traders' Bank Building in Boston, designed by Nathaniel J. Bradlee (c. 1858), and Gridley J. F. Bryant's (at least partly fireproof) Cheshire County courthouse in Keene, New Hampshire (1858–59).[104] More often, as they had in the vault days, New Englanders simply built fireproof rooms in ordinary structures. Thus, Trenton Iron Works's main customers for beams in New England were two firms that made bank vaults and safes: Smith, Felton & Company and Denio & Roberts. Smith, Felton & Company made the ironwork for the San Francisco customhouse.[105] While the firm could build whole fireproof floors—it was the contractor for both the Traders' Bank and the Keene courthouse—most of its fireproof construction business involved installing fireproof rooms. It put these in several Boston buildings: the old Suffolk Bank; a building at State and Congress Streets; and in the offices occupied by Boston's Registry of Deeds.[106] Denio & Roberts, blacksmiths and machinists, expanded into the architectural iron business; an 1857 advertisement for the firm stated that they manufactured "iron store fronts, ironwork for public buildings, columns, girders, and beams."[107] Like Smith, Felton & Company, Denio & Roberts learned about iron and brick construction through contracts it had with the federal government. It supplied the ironwork for the Bristol, Rhode Island customhouse (1856–58).

New kinds of owners—for example, colleges—and owners in the West began to adopt fireproof construction. For a building at Union College in Schenectady, architect William Wollett made the ground floor vaulted and installed an iron and brick fireproof room to house a library and museum. Vassar College, a new women's college in Poughkeepsie, built an iron and brick gashouse and heating plant for the main college building.[108] In 1855, Princeton College in Princeton, New Jersey, rebuilt its fire-damaged Nassau Hall with the iron and brick system. Its architect, John Notman, who built the iron and brick Academy of Natural Sciences in Philadelphia fifteen years before, used railroad rails for beams in this reconstruction. A few iron and brick buildings went up in Western cities in this period: the cities of Detroit and Chicago built fireproof houses for their waterworks' pumping engines.[109]

Architects of the 1850s agreed that they were living in the age of iron, and while many criticized cast iron facades, on aesthetic grounds, few objected to using iron structurally. Indeed, they had little reason to, since the new iron and brick system had little visible impact. The walls of fireproof buildings might have been somewhat thicker than those of an ordinary building, depending on the weight of the floors. But for a building of moderate height, this difference would have hardly been noticeable. A fireproof building could be wrapped in any of the architectural styles of the day. Ammi Young took advantage of this flexibility to discard the Greek revival style, with its costly architectural elements, and adopt the Renaissance palazzo style for federal buildings.[110] Architects could conceal brick arch floors under lath and plaster ceilings; cover iron girders with decorative plaster moldings; and design iron columns in classical shapes and painted in imitation of marble. The internal structure could be made invisible.

While the iron and brick system freed architects to work in a range of styles, it required them to design construction details more precisely. The supervising architect's office occasionally specified the wrong length for columns, which resulted in some creative adaptations of the building design or else ordering new columns. Professional architects designed only a small share of all buildings at the time, but a fireproof building could not be put up without an architect's services, since it required measured drawings.

By the end of the decade, the iron and brick system defined what fireproof construction meant.[111] New York City's first comprehensive building code, in 1860, defined a fireproof building as an iron and brick building. Although considerably more practical than the vault style, it still had drawbacks that discouraged people from adopting it. Brick segmental arches filled with concrete were heavier and took up more headroom than ordinary wood-framed floors. Many owners found the scalloped brick floors crossed with tie-rods unsightly, although they could be covered with a drop ceiling. But a major stumbling block continued to be the high cost of iron and brick construction: by one contemporary estimate, a fireproof building cost more than double a comparable sized one of ordinary construction.[112] Yet the need for better fire protection was as great as ever; indeed, in the latter 1860s, the amount of property lost to fire had reached an alarming level. And the fact that some owners would build iron and brick structures despite the high cost suggested that more might do so if the price was right. In the next decade, inventors and architects worked to improve the ingredients of the fireproof building and bring down their cost.

RESPONSE TO THE
GREAT FIRES

Experimentation in the 1860s
to the Early 1880s

It is but a few years since it was . . . generally accepted . . . that an incombustible building was fire-proof. . . . But experience has demonstrated that such structures, though they will not *burn*, may still be *destroyed*. The new problem that confronts us therefore is, how to preserve the materials of construction from the effects of fire.

PETER B. WIGHT, architect, "The Fire Question," 1877

In the chronicles of urban conflagrations, the Civil War years are a blank. Fire insurance companies dropped customers in the war zones and so had no business reason to track fire losses there. Some Southern cities burned during the war. In the North and West, conflagration, at least, did not add to the trials of that unhappy time.

A construction boom followed the close of the war, yet relatively few of the buildings that went up in the postwar years were fireproof, even though the materials to erect them were more readily available than ever. By the end of the 1860s, practically every large city had at least one foundry capable of making structural castings, and a few iron companies—notably Architectural Iron Works in New York City and Hayward, Bartlett, and Company in Baltimore—served a national market. Several rolling mills produced I-beams, in addition to Trenton and Phoenix Iron Works, and builders could get larger (deeper and longer) beams. By 1868, three mills—Buffalo Union, Phoenix, and Trenton—rolled fifteen-inch deep sections.[1]

Little changed in the clientele and geography of fireproof construction from the 1850s. New York City continued to be the place where most fireproof buildings went up. Government—both federal and local—as well as banks, institutions, and some city-based manufacturers (e.g., publishers) still comprised the main customers for fireproof buildings. Examples of fireproof buildings in New York City from the postwar years include the infamous New York County ("Tweed") Court House (1861–72); A. T. Stewart's Women's Home, later called the Park Avenue Hotel (1869–77); Park National Bank (c. 1868); and the renovated City Bank in Wall Street. One unusual sort of fireproof building was a freight depot, which the Hudson River Railroad Company built on St. John's Park (c. 1867)—perhaps the only fireproof warehouse in New York City.[2] Outside Manhattan, Hupfels Brewery in the Bronx added a fireproof section to its plant (c. 1865) and Kings County (Brooklyn) completed a fireproof courthouse in 1865.[3]

A handful of owners erected fireproof buildings outside New York. In Chicago, a city that had suffered four large fires in the 1850s and 1860s, the Historical Society built a new fireproof home for itself (1866). The state of Illinois put up a fireproof prison in Joliet. Cleveland's city hall (c. 1870) had "floors resting on iron beams with brick arching." In Washington, D.C., James Renwick Jr. designed a fireproof art gallery for William Corcoran (completed in the 1860s). The Smithsonian ordered an iron roof for the Castle to replace one destroyed in an 1865 fire, and at the same time put iron and brick floors in the building's towers. Boston made its new city hall (1861–65) partly fireproof, using iron and brick construction in the first three floors. In Philadelphia, Samuel S. White erected a fireproof building for his artificial teeth and dentistry appliance manufacturing business (1867).[4]

Insurance companies were the only new sort of clients for fireproof construction. One of the earliest of these to put up a fireproof office building (c. 1861) was Continental Insurance Company in New York City. An insurance journalist praised the building with the paradoxical suggestion that "When the other buildings of our great city become as safe from risk of fire as the new Continental building, there will be little need for fire insurance companies."[5] Other fireproof insurance company offices included the Mutual Life Insurance Company building (c. 1863–70), the Columbia Insurance building, and the New York Life Insurance Company building (1868–70)—this last one built on a site cleared by fire in 1867.[6]

But these buildings were exceptional, standing as lonely fortresses amid their combustible neighbors. The public's apparent indifference to adopting fireproof construction perplexed an editorial writer for *The Nation* in 1867. "The American people," he wrote, "adopt with great readiness any new method of accomplishing an object of practical utility," expanding the street railways at a great rate, using sewing machines and the telegraph. Unaccountably, they avoided some valuable technologies; of these, neglecting to adopt the means to protect "valuable buildings and their contents from destruction by fire" was the least excusable. Instead of spending money to recover from fires, he argued, Americans should make their buildings less combustible in the first place.[7]

Some Alternatives to Brick Arches

Architects and inventors introduced alternatives to brick arches, to overcome the various drawbacks of that system. One new kind of floor, similar to the kind used in small industrial buildings and warehouses in Great Britain, substituted stone slabs for brick arches. The flat stones made a thinner floor and, because they rested on the bottom flange of the beams, created a flat ceiling that could be plastered directly. Such a floor could go up faster than brick arches. Several architects proposed stone floors in the 1840s and 1850s for buildings they designed, although none went forward. In the 1860s, these floors appeared in a few buildings: the American Exchange Bank and Continental Bank buildings in New York City, and the Mutual Benefit Life Insurance Building in Newark, New Jersey. In one of these, the masons finished the underside of the stone slabs "with tracery patterns to form an ornamental ceiling."[8] Whatever its structural and practical advantages, this system must have cost more than a brick arch floor.

Another fireproof floor system introduced in this decade used corrugated iron sheets and concrete, which made lighter floors that could be installed faster than brick arches. J. B. and W. W. Cornell, important iron founders in New York City, built such floors in the 1850s, for example, in the Bank of the State of New York (1856–58) in New York City (Fig. 3-1).[9] An engineer who watched this building being demolished at the turn of the century left a detailed description of the floor system. It consisted of eighteen-inch-deep girders and ten-inch beams spaced five-and-a-half feet on center, supported on the bottom flanges of the girders; both the beams and girders were made from rolled iron plates. The beams carried deeply corrugated iron sheets, with corrugations measuring about a foot apart and six inches from top to trough, which were covered with concrete to make a level walking surface. Other New York buildings with these floors included the Columbia Insurance building, Bank for Savings on Chambers Street, and the Fulton Bank. Despite their advantages compared with brick floors, corrugated iron floors cost more. Brick floors continued to be the most widely used kind in fireproof buildings.[10]

Isaiah Rogers, then serving as supervising architect of the Treasury Department, employed a variation of this corrugated iron system to make the

Fig. 3-1. Advertisement showing the various iron products made at J. B. and W. W. Cornell's Iron Works, including plate girders, window shutters, and stairways as well as a specialty: a corrugated iron and concrete deck used for floors and roofs. David Bigelow, *History of Prominent Mercantile and Manufacturing Firms* (Boston, 1857).

roof of the west wing of the Treasury Department building in Washington (1862–63). It consisted of trusses—of an unusual lattice type—spaced roughly five-and-a-half feet apart; the trusses carried arched iron sheets on both the bottom flanges and at the top. Covered with concrete, the bottom plates would have prevented a fire inside the building from burning through the roof, while the top plates made a barrier against fire from the outside.[11] The Treasury Department used this roof system in other buildings it erected in the late 1860s and early 1870s, including the north wing of the Treasury building and the post office/federal courthouse in New York City.

Joseph Gilbert, a Philadelphia builder, introduced another variation on the iron deck idea. In his 1867 patent, Gilbert proposed installing iron plates under wooden floors and roofs to shield them from fire, similar to an approach used by Robert Smirke some forty years earlier in the King's Library at the British Museum and the General Post Office in London. Gilbert's plates were corrugated and either flat or curved. Cast iron bearers, scalloped like the edges of the plates and attached to the lower flanges of the beams, held the plates in place. Gilbert claimed his "cheap, light, and ornamental fire-proof ceiling" allowed for longer spans than was usual for brick arches, and the underside did not require plastering.[12] He estimated that it would cost 25 percent less than brick floors.

Gilbert's patent described a second form of floor, in which concrete completely filled the space over the iron plate to the tops of the beams (Fig. 3-2). In this case, the plates served as permanent centers for a load-bearing, plain concrete floor. This idea also had precedents, most notably one illustrated in William Fairbairn's well-known 1854 book on fireproof construction. Fairbairn does not mention any extant examples of the type, however, so it may have been only a suggestion.[13] Because of the floor's novelty, Gilbert solicited endorsements of it for his advertising. He submitted a sample to the Franklin Institute in Philadelphia, where an examining committee concluded that the floor met Gilbert's claims and was as fireproof as alternatives. A group of Philadelphia architects seconded this committee's report. Gilbert also subjected the floor to a load test, conducted for a group of witnesses that included John C. Trautwine Jr., author of a leading engineering textbook. The arch in the test was made of an 18-gauge iron sheet, twenty-eight inches wide, which spanned six feet between supports. It successfully carried six tons.[14]

Unlike the other systems, this type of concrete floor became relatively popular in the late 1860s through the 1880s. Most of the fireproof build-

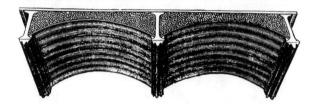

Fig. 3-2. Cross-section through a floor consisting of a corrugated iron arch filled with concrete, patented by Joseph Gilbert in 1867. This kind of floor continued to be used through the end of the nineteenth century. *Architectural Review and American Builders' Journal* 1 (July 1868).

ings erected in Chicago before 1871 had these floors, including the Chicago Tribune Building (1868–69); the First National Bank (c. 1868); the new wings of the county courthouse; the Nixon Building (c. 1871); and the ground floor strong room of Fidelity Safe Deposit Company. The city's post office (1873–80), erected after the Great Fire, also had iron and concrete floors.[15] The Union Foundry Works in Chicago made these floors under license from Gilbert.[16] In New York City, the Equitable Life Assurance Company adopted this system in its new headquarters, constructed between 1868 and 1870. This company also used the system in its Boston branch office (1873–77), which was a ten-story building.[17] In San Francisco, the new city hall, started in the early 1870s, contained corrugated arch floors in some sections.[18]

One last alternative to brick arches from around this time was a floor made of large, clay blocks, installed in the Cooper Institute (now Cooper Union) building in New York City. Patented by the building's architect, the German immigrant Frederick A. Peterson, this floor utilized the same material as the French and British floors made of pots, tubes, and hollow blocks.[19] Unlike these, Peterson's floor consisted of a single large block that spanned about two-and-a-half feet between beams. The block had a flat bottom and arched top, and rested on the bottom flanges of the beams; the top was leveled up with concrete.[20] Since they were made by hand, and, given their large size, with difficulty, Peterson used them sparingly; only the floors over the auditorium contained the blocks. No other architect adopted Peterson's tile.

By the 1860s, Americans had developed methods of fireproof construction that equalled those of Great Britain and France. The 1867 Universal Exposition in Paris allowed them to compare their methods with those of the Europeans. While the commissioner who reported on the building materials and methods presented there admired France's large rolled girders and concrete construction, he believed noncombustible building materials as good as those in Paris were available in New York. But unlike New Yorkers, Parisians actually used such materials routinely, which he believed accounted for the comparative freedom from destructive fires and "remarkably cheap" insurance in Paris.[21] Another commissioner, ironmaster Abram Hewitt, found nothing remarkable in European iron construction. He concluded, "Thus far the construction of a fire-proof building in the United States is accomplished with less pounds of iron for a given strain per square foot than in France, and we have nothing to learn from the Exposition in this respect."[22]

Yet British architects had become dissatisfied with iron and brick buildings, in part because they considered them unreliable in fires. An 1861 conflagration in a London warehouse district—the Tooley Street fire—confirmed their doubts. To compound the disaster, James Braidwood, head of London's Fire-Brigade, died fighting this blaze, crushed by a wall that collapsed on him. The fire started in and spread among the very sort of iron warehouses Braidwood had warned against. His successor, the redoubtable Captain Eyre M. Shaw, likewise objected to iron construction. Shaw's idea of a fire-resistant building was one constructed with hardwood posts, wood girders, and joists, with cement filling the space between the joists. An iron frame might fail suddenly—something firefighters greatly feared— whereas a timber one would give some warning. He condemned stone, the material most commonly used in London stairs, because of its tendency to fracture in a fire. Shaw dismissed fireproof construction: his handbook on fire protection, *Fire Surveys*, did not even discuss it.[23] His prescription for reducing fire hazard was to limit the size of buildings—cubical capacity as well as height—rather than attempt to make them "fireproof."

Revered for his heroism, even immortalized in a song in Gilbert and Sullivan's *Iolanthe*, Shaw wielded great influence, and architectural writers cited his views for years to come.[24] For example, one of these, the architect T. Hayter Lewis, wrote in 1865, "Nothing short of diminishing the size of warehouses and other such buildings, or protecting them with brick arches or brick piers [as in solid masonry buildings], will render them secure." Indeed, he continued, "the common method of brick arches on iron girders is the most dangerous that can be used."[25] Consonant with Shaw's recommendations, London's building laws limited the size of buildings and did not require any sort of building to be fireproof. A distrust of iron also led British architects to adopt concrete in place of iron and brick arches, and to even avoid using iron to reinforce the concrete.

This trend away from iron led one British architect to exclaim, "What changes have not come over the meaning of fire-proof! Among my early experiences no building was considered fire-proof that had not iron joists and brick arches. . . . Now it is generally admitted that no floor can be called . . . fire-proof that has iron used in its construction—such is the present position of iron." He rejected the idea that a building could be fireproof, and suggested replacing the term with *fire resisting*. If a builder did use iron structurally, he should encase it, to protect it from heat.[26] Despite such admonishments, British mill designers continued to use the iron and brick fireproof system, and they rarely covered structural iron.

Some American architectural writers adopted these opinions about iron. That fire could destroy structural iron, one wrote, proved that "iron is unreliable as a building material" and therefore not a fit ingredient for a fireproof building. Echoing Captain Shaw, this writer suggested that wood could be "rendered incombustible at no great expense" and recommended concrete floor systems such as were used in Britain, although they were unavailable in the United States.[27] American engineers, too, were wary of structural iron, mainly because it expanded when heated and rusted. Many members of the audience at a meeting of the Polytechnic Club in New York City, after a presentation about cast iron building fronts, raised concerns about using iron as a building material, although one acknowledged little was known about the consequences of expansion.[28]

And indeed, so few iron and brick buildings existed in the United States that the fire resistance of the system had not been tested.[29] One of the first real tests occurred in 1866, when Ammi Young's 1854–58 customhouse in Portland, Maine, stood in the path of a conflagration that destroyed the city center. Portland had a reputation for being a substantially built city, although individual structures did burn; in fact, Young's customhouse replaced the previous one, which burned in 1854. Still, the downtown contained conditions fertile for conflagration, including boat and lumber yards. Youths setting off firecrackers probably started the July 4 fire; with townspeople off celebrating, it spread unchecked and soon turned Portland's business district into a "perfect sea of flames." Yet the customhouse remained intact: it even sheltered a claims agent trapped inside.[30] The fire seriously damaged the facade, however, and for this reason, the Treasury Department ordered the building demolished. No one seems to have studied the event for the lessons it might teach about the effect of fire on a fire-proof building.[31]

But the question of the performance of fireproof buildings in fires soon became a preoccupation for a group of architects and engineers, the nation's first fire protection experts. The great fires in Chicago and Boston encouraged some men to make structural fire protection a specialty. Also, at the time of these fires, the professional infrastructure existed to support the development of this new discipline. After the Civil War, both the national organizations for architects, the American Institute of Architects (AIA), and for civil engineers, the American Society of Civil Engineers (ASCE), began to hold regular meetings and publish proceedings.[32] In 1876, America's first long-lived architecture periodical with a national circulation, the weekly *American Architect and Building News* (AABN), began publication.[33] These venues for exchanging information enabled a group of designers to develop the field of structural fire protection.

The Great Fires at Chicago and Boston

The Great Fire of 1871 in Chicago provided the first extensive field test of iron and brick fireproof construction: this time, a number of architects examined the ruins to understand why the fireproof buildings failed and how they could be improved. The fire leveled a huge portion of the city and, by a contemporary estimate, left roughly one hundred thousand people homeless, yet it began as just one more blaze in a city that had experienced four sweeping fires in recent years. Among the factors that contributed to making this fire the disaster it became was a severe drought. So little rain had fallen in the region—only two-and-a-half inches in Chicago itself during the summer and fall—that this wooden city was tinder dry. On the same day Chicago burned, a forest fire struck a similarly parched lumbering area to the north, near the town of Peshtigo, in Wisconsin. The Peshtigo fire killed over a thousand people, a much greater loss of life than in Chicago, making it one of the deadliest fires in U.S. history, while the Chicago fire was one of the costliest.[34]

Contrary to ill-informed reports from the time, the fireproof buildings performed creditably in the fire.[35] Ammi Young's post office resisted the attack of the fire for a long time. An eyewitness noted that it served as a barrier, preventing the fire from passing. But the flames spread north along a nearby street, then turned east, then down "the south line of the street like a whirlwind and, turning Dearborn, melted away Reynolds

block . . . , [which brought] them to the north side of the post office";
meanwhile, another column of fire swept toward the post office from the
west. "Before this joint attack, [the post office] yielded. . . . Although its
walls stood bravely, its interior was soon gutted."[36] But enough of Chicago's
other fireproof buildings survived so that they could be reused, whereas
the ordinary buildings had been reduced to heaps of brick and ash. Parts of
the fireproof First National Bank, courthouse wings, the Nixon block, and
the Fidelity Safe Deposit strong room remained standing. In fact, the fireproof Tribune Building stood largely intact: its south wall bulged in, and a
section of its Madison Street facade collapsed when a heavy safe fell from
an upper floor, pulling down the floors and wall as it descended; yet the
building was reused after repairs.[37]

The Chicago fire occurred a few weeks before the annual convention of
the AIA and naturally was the talk of the meeting. Peter B. Wight, an architect who became an early expert on fireproof construction, reported on
his personal inspection of the burned district and observations about the
performance of the fireproof buildings. He blamed the failure of cast iron
columns, which he believed broke in the fire and let down the floors above,
for causing the destruction of the "so-called fire-proof buildings." Yet he
noticed that the lower flanges of beams exposed to fire sagged, and therefore suggested that it "may be proper to cover beams with cement, artificial stone, or terra cotta, or plaster."[38] In this seeming afterthought, Wright
articulated a key idea: he realized, although tentatively at this point, that
to survive a fire, buildings could not be merely noncombustible, but also
had to resist the effects of heat. The next phase of fireproof construction
technology developed from this insight.

Wight also criticized the shoddy practices of Chicago's jerrybuilders for
abetting the fire, and he believed that destruction on the scale of Chicago's
would be impossible in well-built cities like New York or Boston. But just a
year later, in another conjunction of unfavorable circumstances, the "Fire-
fiend" attacked the "colossal business structures" of Boston.[39] Whereas
tinder-dry conditions helped spread the fire in Chicago, a popular style of
roof—the mansard—served as the culprit in Boston's fire. The "French
roof" allowed more usable space in attics compared with a pitched roof.
Yet the steeply sloping sides of the roof were simply wooden walls covered
with roofing material, and were made even more combustible with wood
trim around the windows and wood cornices. Boston's inspector of buildings warned about these roofs in 1871, complaining that while they were
"dangerous in the extreme," practically every building then under con-

struction in Boston had one.[40] He proposed restricting them; but by then, it was too late. Most observers believed the fire spread from building to building along the mansard roofs. Mercifully, this fire did not spread beyond the commercial center of town and so destroyed few homes. Nevertheless, the area it destroyed contained the most valuable property in the city: by one estimate, the fire wiped out 8 percent of the total capital in Massachusetts.[41] Unlike Chicago's, Boston's conflagration shed no new light on the performance of fireproof buildings because none of Boston's iron and brick buildings stood within the burned district. The city's new iron and brick post office, still under construction at the time, not only survived but also barred the fire from spreading west.

Nevertheless, at the AIA meeting of 1872, architects tried to draw lessons about building materials and fire protection from Boston's fire. Richard M. Hunt proposed that owners install sprinklers in their roofs, like those in factories at Lowell, Massachusetts. Another architect recommended owners install standpipes with hose outlets in their buildings for fire fighting, like the owners of Lord & Taylor had done in their new store (1869–70) in New York City and the American Express Company (c. 1873) did in their Chicago building. Robert G. Hatfield, an architect who specialized in the technical aspects of construction, was especially interested in the effects of fire on iron. In contrast to Wight, Hatfield found that while in Boston, as in Chicago, "all iron . . . proved untrustworthy; cast iron, however, had stood much better than wrought iron."[42] Wrought iron beams, in other words, were more vulnerable to failure than cast iron columns.

Architects gave papers on fire protection at AIA chapter meetings as well. In 1873, Nathaniel H. Hutton, an architect and engineer, presented a paper on fireproof construction to the Baltimore chapter. Cincinnati chapter members heard papers on the prevention of fires and on state building laws. The Philadelphia chapter discussed fireproof construction, the Boston fire, mansard roofs, and the necessity of a new building law for that city. Boston architects naturally focused on fireproofing and building law at the meetings of the Boston Society of Architects.[43] Architects' interest in the topic persisted throughout the decade: A. J. Bloor, Robert Hatfield, Detlef Lienau, Franz Schumann, and Peter Wight all read papers on fireproof construction at AIA annual meetings.[44]

While this emerging fire protection fraternity did not always draw the same conclusions from the great fires, they did agree on two points: first, that brick held up far better than stone in a fire, and second, that iron, while noncombustible, could not be considered fireproof. The Boston conflagration, in particular, highlighted the weaknesses of many kinds of building stones, which were liable to spall, crack, or disintegrate in a fire, especially after being pounded with high-pressure water. Brick walls, on the other hand, "if well constructed, had given up their contents uninjured."[45] Up to this time, architects left iron columns and beams exposed. Indeed, some treated exposed iron beams—which revealed the mechanics of a building's frame—as an architectural feature. One of these, the architect Leopold Eidlitz, praised his colleagues in New York for not hiding structural iron, but displaying it "in broad daylight as an essential feature and preponderating element of modern construction."[46] Unfortunately for those who felt like Eidlitz, iron had to be hidden if it was to be insulated from fire. The solution to the "fireproofing" problem, as the architect A. J. Bloor termed it, was to protect structural members with materials that were "non-conductors of heat, as well as non-combustible."[47]

A Decade of Development: New Fireproofing Materials in the 1870s

The 1870s saw an outpouring of inventions and experiments aimed at improving methods of fireproof construction. Although an economic depression that began in 1873 slowed building construction, and consequently the adoption of new materials, the products introduced in this decade eventually developed into two new systems of fire-resistive construction. The first of these involved hollow building blocks, which replaced brick floor arches and partition walls. The hollow blocks were fabricated from concrete or, more commonly, clay. The second system represented an evolution of the strategy of employing fire-resisting materials to protect an ordinary wood structure. Manufacturers devised protective barriers made out of clay and concrete plates, and also plaster on iron lath. Sometimes described as "fireproof," buildings with protected timber frames were more correctly styled "semifireproof." The semifireproof system appealed to many owners because it seemed to offer as much security as fireproof construction at a much lower cost.

Fired clay, called either "terra cotta" or "tile," was the most important new construction material of the 1870s. While the terms *terra cotta* and *tile* were synonyms, contemporaries always used the former term when referring to ornamental and exterior building materials (e.g., blocks in building facades). They called clay used for fireproofing and for nonornamental, structural blocks (hollow blocks) either tile or terra cotta. To try to avoid confusion, in this text the word *tile* will be used when discussing the latter kind of product.

The idea of lightening fireproof floors by employing hollow tile blocks was a rediscovery of a practice used in ancient Rome. At the end of the eighteenth century, French architects set flowerpot-shaped clay units in plaster to make fireproof floors and roofs. English designers noticed these pot spans on visits to France, and by the 1790s, at least one English manufacturer produced "arch pots." John Soane put them in the dome in his Bank of England Stock Office, and the textile manufacturer William Strutt used them in ceilings over the top floors of two early fireproof factories. The pots were handmade at first, but in the 1830s, at least one English manufacturer employed a machine to shape the clay.[48]

In the 1850s, a variation on the pot idea—hollow tile blocks—appeared in England. One early example, patented by architect Henry Roberts in 1849, consisted of blocks shaped like small, hollow wedges that lacked interior walls (called webs). Roberts used the blocks to build the floors and roofs of model tenement houses he designed for Streatham Street, Bloomsbury, and thereby rendered them fireproof. A model lodging house in Portpool Lane, designed by Roberts and built with his hollow brick floors, survived a serious fire. Roberts's system received much publicity during the Great Exhibition of 1851 in London.[49] Another early hollow block floor was installed by the manufacturer Titus Salt in his huge textile mill, Saltaire (1850–53), located near Bradford in England, in order to protect the mill against fire. His hollow block floors weighed less than brick arches yet provided "security in every room of the factory."[50] A few years later, in 1858, a clay manufacturer, Joseph Bunnett, patented a hollow block floor similar to the one in Saltaire. Bunnett's blocks were about the size of large bricks, and in fact he used brickmaking machines to form them. A notable advantage of Bunnett's blocks was that they could be used for long spans, in either a curving or nearly flat arch.[51] He proposed two styles of blocks: one with the cavity running perpendicular to the supports and the other with the cavity running parallel. While Roberts's and Bunnett's systems, as well as the floors at Saltaire, received favorable notice, they were little used. By the 1860s and 1870s, English owners had turned to concrete systems to make fireproof floors.

Parisian designers, in contrast, used hollow tile extensively in public buildings since the eighteenth century, and in the 1860s, made it the standard material for floors in residential buildings.[52] At this time, at least one manufacturer produced blocks in the shape of voussoirs for making floors. The French hollow tile floors weighed less than both British tile floors and French concrete floors, and they became the model for the first successful American hollow tile floors.

Before the 1870s, with the exception of Peterson's floors in the Cooper Institute, clay had not been used in the United States in hollow structural units or for fireproofing. Rather, hollow tile systems developed simultaneously with the growth of America's clay products industry. At the end of the 1860s, a few firms began to produce ornamental items in terra cotta, such as garden decorations and statues, copying the fashion of Great Britain. In the next decade, the structural tile/fireproofing industry took shape. Firms specialized in making either architectural terra cotta—such as glazed or molded blocks for walls, decorative chimney tops, finials, and cornices—or else materials for what came to be known as the fireproof building or fireproofing industry. The latter companies made blocks for constructing floors, partitions, and roofs, and for covering ("fireproofing") structural metal and wood frames.

The pioneer firms in the fireproofing industry began operations at roughly the same time, in 1872 or 1873. Two of these started in New York City and one in Chicago, founded by a transplanted New Yorker. Later in the century, both cities claimed to be the birthplace of hollow tile, which was a critical ingredient in the early skyscrapers. New York partisans, such as William J. Fryer Jr., a chronicler of New York's commercial development, wrote that New York City had the first hollow tile building. Chicago's Peter Wight, an architect who started his career in New York and became an authority on fireproof construction, claimed that Chicago had the first one.[53] In fact, the men who started the industry knew each other and got their ideas from the same sources: French and British precedents.

The history of hollow tile floor blocks begins with a large tile, similar to Peterson's, patented by the New Yorkers George H. Johnson and Balthasar Kreischer in early 1871.[54] Both men had been involved with iron and brick fireproof construction in New York since the system's early days. Johnson, who had apprenticed in the building trades in his native England before emigrating, worked in the 1850s as a designer at Daniel Badger's Architectural Iron Works. After the Civil War, he practiced architecture on his own for a time, in New York and in several other cities, mainly designing cast iron buildings and, at the end of the decade, fireproof grain elevators.[55] German-born Kreischer started a brick business in 1845, where he manufactured various brick and tile products including fire-brick, a kind used to line furnaces and fireplaces. Kreischer supplied fire-brick for the furnaces at the fireproof Assay Office in New York City. Exactly how Johnson and Kreischer got together for the joint patent is unclear, but their collaboration—if that is what it was—soon ended. On the same day they applied for their patent, Kreischer put in an application for another floor tile in his name alone, one in the same shape but made of three blocks.[56] Probably neither of these was ever produced.

After the Chicago fire, Johnson joined the flock of architects and building tradesmen who went to that city in hopes of cashing in on the post-fire rebuilding boom, where he assumed he would find a market for his tile floor blocks. Joseph Bunnett, thinking likewise, sent a sample of his hollow tile arch floor from England, and it was erected on a vacant lot outside the office of one of Chicago's busiest architects[57] (Fig. 3-3). But owners had no time to investigate new materials and generally stuck with familiar and readily available ones. Nevertheless, Johnson received several contracts for his blocks, a remarkable accomplishment considering that they had never been used before.

The Kendall Building (1872–73), designed by Johnson and John M. Van Osdel, was the first building to contain Johnson's hollow tiles. Interestingly, the floor blocks, in the form of a flat arch, were unlike those of the 1871 patents, but rather resembled types manufactured in France. Perhaps

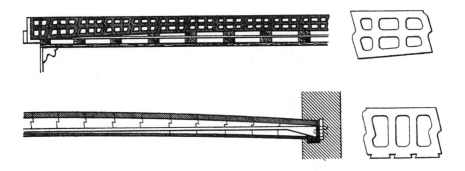

Fig. 3-3. Joseph Bunnett's hollow tile floors. *Above*, a shallow arch made of blocks with webs running parallel with the supports. *Below*, an early example of a form that came to be known as "end-construction," in which the webs of the blocks run perpendicular with the supports. John J. Webster, "Fire-proof Construction," *Minutes of Proceedings of the Institution of Civil Engineers* 12 (1890–91).

Johnson got the idea for the Kendall floor from one designed by Vincent Garcin, patented in France in 1867. Another source may have been the Roux Brothers, a French firm that manufactured tile floor blocks. A circular from this firm, printed in 1868, which illustrated voussoir-shaped flat floor arches similar to Johnson's, had made its way to American architects by the early 1870s (Fig. 3-4). Johnson patented his new system in 1872. The Kendall also contained hollow tile partitions, an item Johnson patented in 1871.[58] Locals immediately appreciated the importance of this building: "a party of gentlemen prominent in city affairs and local history" came to observe its construction.[59] Johnson's two other fireproofing projects in Chicago were the (third) Palmer House Hotel, completed in 1875, and the Criminal Courts Building and Jail, completed in 1873. A few years later, the successor to Johnson's company made hollow tile floors for the Cook County courthouse and adjoining city hall (built slowly between about 1877 and 1885).[60]

The hollow tile industry in New York got off the ground when Kreischer licensed a fledgling engineering and architectural iron firm, Heuvelman, Haven & Company (HHC), to produce floors under his patent.[61] HHC's principals were veteran architectural ironmakers: Wilson Haven had worked with James Bogardus and the Architectural Iron Works, and then for Novelty Works in Brooklyn, where John Heuvelman had been a draftsman. These two worked for Novelty when it had contracts to make iron for two prominent fireproof buildings: the New York City post office and the Williamsburgh Savings Bank in Brooklyn. After Novelty fell apart, HHC took over some of its contracts.[62] The partners got involved with hollow tile through their experience working on fireproof buildings.

In early 1873, HHC offered hollow tile blocks to A. B. Mullett, supervising architect of the U.S. Treasury Department, for use in the New York post office, then under construction. While it listed "hollow fire-proof tiles for floors, partitions, and vault linings" among its products, apparently HHC had not manufactured any tiles yet, since the partners wrote to Mullett that "in a short time [we] will have specimens to exhibit." Because they did not expect Mullett to be familiar with hollow tile, they assured him the material was "not a new thing but is used in various forms in all first class prominent Buildings in Paris and other principal Cities in Europe." Mullett made a contract with HHC, and thus the floors of the halls as well as several rooms in the upper floors of New York City's post office (1869–75) were built with hollow tile.[63]

From drawing attached to Roux Frères Fr. Patent: Mar. 25.1868.

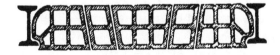

From Circular issued by Roux Frères. 1868.

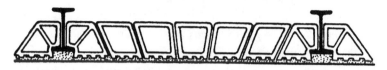

Terra-cotta Arch in Kendall (Equitable) Building, Chicago.

Fig. 3-4. Various patterns of hollow tile floor blocks made by the Roux Freres, France, 1868 (*top three*) and a floor arch modeled on the French blocks used in the Kendall Building, Chicago, 1872 (*bottom*). Peter Wight, *The Brickbuilder* 6 (April 1897), and J. K. Freitag, *Fireproofing of Steel Buildings.*

In the same year, a third company entered the fireproofing business: the Fire-proof Building Company of New York (FBC), established by Leonard F. Beckwith, a civil engineer and architect who had been educated in France. Beckwith served as a commissioner to the Paris Universal Exposition in 1867 and wrote a report on the structural concrete (called *béton-coignet*) exhibited there and in several buildings he visited in France.[64] He foresaw a market for this material, which could be shaped *en masse* (cast in place) or formed into building blocks, in America and started a company to manufacture concrete blocks for arch floors and partitions in patterns similar to those used by HHC. Beckwith's formula for concrete included French cement, plaster, and coal cinders. While he apparently was the first American to use this combination of ingredients, surprisingly he did not patent it. Perhaps the material had already been patented in France. Kreischer held a patent for concrete, from 1871, which he intended to use

to make hollow blocks; although similar to Beckwith's, Kreischer's contained fire clay instead of French plaster.[65] Whether Kreischer ever made
concrete blocks using his patented recipe is unknown.

FBC also made hollow blocks of tile, as well as concrete, under a license
from Johnson. By 1874, the firm held its own patent for a flat tile arch, one
invented by Arthur Beckwith—a civil engineer and architect, member of
the company, and presumably Leonard's brother.[66] Exactly why the firm
used clay in some cases, and concrete in others, is unclear. In one early project, the American Museum of Natural History in New York (first section
built 1873–77), FBC installed concrete floor blocks. The Peabody Institute
of Baltimore ordered concrete floor blocks for its new library, which FBC
made and installed in the winter of 1876–77.[67] In other buildings, Beckwith built clay floor arches, for example, the Delaware & Hudson Canal
Company's Coal and Iron Exchange (1873–76) and the Tribune Building
(1873–75), both in New York City.[68] Peter Wight, the fireproof building
expert, suggested that concrete blocks were inferior to the tile, and only
after Beckwith "substituted burned-clay hollow blocks for the plaster
blocks" did he do "any work that was successful and permanent."[69] But
Wight does not explain in what way clay was better than concrete, and he
may have been expressing his own preference rather than those of FBC's
customers. Beckwith continued to manufacture "Hydraulic Lime of Teil
(France)" blocks, along with "Hollow Bricks of all kinds," through the
1870s. He even exhibited his "Teil lime composition" blocks at the 1876
Centennial Exhibition in Philadelphia, for which he received a commen-
dation and a favorable review by the architect Richard M. Hunt, who
wrote a report on building materials at the exhibition.[70]

While it is difficult to trace the early spread of hollow tile outside of
New York City and Chicago, a few examples have come to light. The upper
floors in Illinois' new state capitol in Springfield (installed in the early
1870s) were made of cinder and plaster blocks designed by the architect
Alfred H. Piquenard, who helped plan and build the capitol and had seen
such blocks in France. In St. Louis, the Singer Manufacturing Company's
new fireproof building (c. 1878) contained hollow tile floor arches and
partitions.[71] Also in St. Louis, the new fireproof Southern Hotel (1879–81)
went up to replace its predecessor, destroyed in an 1877 fire, which had
partitions made of "gypsum, sand, cement, and pulverized coke" and
floors of "railroad iron" and "cement." Whether the floors were concrete
slabs, which would have made this one of the earliest examples of a reinforced concrete floor in the United States, or cement blocks, which is how

they were described in a contemporary account, cannot be determined. The hotel's owner felt such confidence in its "infallibility that he will not insure the house for a dollar."[72] A notable office building from the 1870s, Milwaukee's Mitchell Building (completed in 1878), was a showcase of hollow tile. It featured hollow tile skewbacks in its floor arches; cruciform fireproof columns protected with tile cladding; hollow tile partitions; and a roof of tiles set in T-iron roof rafters. Not surprisingly, the Mitchell was also "the most costly private building erected in the West for many years."[73]

Hollow block floors had the disadvantage of costing more than brick arches or even Gilbert-style iron arches. Unlike the latter two, hollow tile floors were custom-made, designed to fit the spacing and depth of the beams, and contained expensive materials such as cement, plaster, and fire clay.[74] In their favor, tile arches weighed less than iron or brick arches, and therefore a designer could economize on material in the other load-bearing parts of the structure to offset the extra cost of the tile arches. Nevertheless, brick arches remained the top choice for fireproof floors in the 1870s because they were "the simplest, strongest, and cheapest" kind.[75]

Americans may also have been reluctant to adopt hollow tile floors because they did not know how well they would perform: could they carry the required loads and survive a fire? Manufacturers published testimonials from construction experts and satisfied customers, as well as the results of successful tests, to reassure potential customers. HHC conducted a test of its tile floor for the federal government, which they claimed proved the product's "strength and practicability."[76] The Fire-proof Building Company stated in an 1878 advertisement that "Our Fire-Proof work is the only one approved by the New York Board of Fire Underwriters, and the Superintendent of Buildings, after severe public tests."[77] In its first trade catalogue (1882), New York Terra Cotta Lumber Company included reports of tests it had commissioned. One was a series of load tests observed by the well-known engineer, Alfred P. Boller; in another, the commercial testing laboratory of the Riehle Brothers in Philadelphia crushed small cubes of the company's special porous clay to determine its strength.[78]

While such tests satisfied some customers for some purposes, they had the drawback of being idiosyncratic. Each manufacturer might test his product differently, so the results from the various tests could not be compared. Manufacturers naturally did not publicize failures or poor results. In the 1850s, the Treasury Department and Captain Meigs at the U.S. Capitol sponsored tests that produced much-needed information on wrought iron beams; however, the supervising architects of the 1860s and early 1870s discontinued materials testing. One architect lamented the loss of government-sponsored tests. How, he asked, were architects to know the relative merits of new products since they lacked the means to make their own tests, and manufacturers would not provide complete information?[79]

Architects simply dealt with the companies they trusted, and for this reason, the Fire-proof Building Company became the most successful one in the 1870s. Peter Wight recalled that Beckwith's blocks "received great favor from architects at the seaboard cities, mainly on account of the confidence reposed in [Beckwith's] scientific attainments."[80] He described Beckwith as "the first practical expert in fire-proofing in America who put his inventions into extensive use. . . . Mr. Beckwith made his own specifications, which were acceptable to the eastern architects, and carried out his own contracts."[81] Beckwith worked on important structures in many different cities. In New York City, his projects included the Bennett and Drexel buildings (begun 1872); Western Union (begun 1873); Delaware & Hudson Canal Company's Coal and Iron Exchange Building; Tribune Building; and Lenox Library (begun 1871); and the Vanderbilt residence. The Delaware & Hudson Canal Company Building won praise from the president of the National Board of Fire Underwriters, who considered it an exemplary fireproof building.[82]

Although only three companies actually entered the hollow tile business in the 1870s, inventors churned out a stream of ideas for new styles of arch blocks and materials for making them, some of which eventually were adopted by tilemakers. In 1872, Wear L. Drake, who lived near Chicago, patented a flat arch floor system in which the end blocks, which pressed against the beam, wrapped over the top and bottom of the beam flanges, insulating it from heat.[83] In the 1880s, tilemakers began to produce floor systems with blocks that protected beam flanges. Another idea introduced in the 1870s was an "end-construction" style of arch, in which the walls of the blocks ran perpendicular to the beam. The arch blocks used in the 1870s were the "side-construction" type, with walls running parallel to the beams. Probably the first American patent for an end-construction arch went to Arthur Beckwith in 1874.[84] In the 1890s, many tile companies added end-construction arches to their product lines, and the earliest styles resembled Beckwith's model.

Inventors also tried to simplify the design and manufacture of arch blocks, in order to bring down their cost. In 1874, George Johnson patented a type of floor that used just two simple, alternating blocks: clay tubes and concave "binders." Engineers working on a new capitol building for New York State tested a small sample of this floor system (four feet across, one foot wide); although it lacked the properties of an arch, the test showed it could support a heavy load.[85] This particular model probably was little used, but it represented an early attempt to standardize and simplify floor blocks. Another invention along these lines, patented in 1879 by Leonard Beckwith, consisted of only two simple blocks in an end-construction arch. This was a structurally advanced design, yet few architects adopted it.[86]

Another application for tile introduced in the 1870s was blocks for sheathing roofs. Roofs had long been a weak point in fireproof buildings. The iron roofs on many of the Treasury Department buildings of the 1850s soon began to leak and had to be replaced. As early as 1873, George Johnson and William J. Fryer Jr. patented a roof tile in the form of a hollow, rectangular slab, rabbeted on two sides so the blocks would interlock.[87] One of the first buildings to have a roof deck made of hollow tile was the pump house of the Chicago Water Works, rebuilt after the 1871 conflagration. The deck consisted of a framework of T-irons that carried tile blocks; it was covered on the outside with slate. Such roofs became popular, especially in cities where the laws mandated fireproof roofs on tall buildings (Fig. 3-5). To thoroughly protect this kind of roof, designers covered the bottom of the iron framework with tile and plaster.[88]

Porous terra cotta, or porous tile, another important new material, was introduced as a substitute for brittle, solid clay tile products. Invented by the architect Sanford E. Loring in 1874, it was made by mixing clay with sawdust or other vegetable matter, which burned out in the firing.[89] Loring made structural tile products at his Chicago Terra Cotta Company, which was unusual in that it manufactured structural hollow tile and architectural terra cotta.[90] Besides being tougher than solid tile, porous tile was lighter, conducted heat less well and so insulated better, and was better able to withstand warping or breaking when fired. Eventually manufacturers used various recipes for porous terra cotta, aimed at making it even tougher.

The first application of porous tile was in blocks designed to fireproof iron columns. Cast iron columns, the most widely used kind, had a long-standing reputation for being unreliable in a fire. Contemporary writers regularly warned that heat caused them to lose strength and expand, and that they would crack after being heated and doused with water.[91] As early as 1860, New York City's building law described two methods for fireproofing columns. One called for wrapping a column with metal lath and

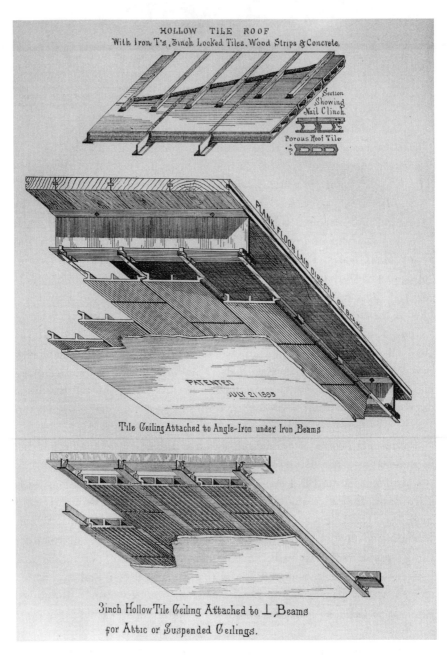

HOLLOW TILE ROOF
With Iron T's, 3inch Locked Tiles, Wood Strips & Concrete.

Section Showing Nail Clinch.

Porous Roof Tile

PLANK FLOOR LAID DIRECTLY ON BEAMS

PATENTED
JULY 21 1885

Tile Ceiling Attached to Angle-Iron under Iron Beams

3inch Hollow Tile Ceiling Attached to ⊥ Beams
for Attic or Suspended Ceilings.

Fig. 3-5. *Top,* "book tile" type of hollow tile roof decking, carried on iron Ts. The deck was finished with wood nailing strips, for attaching the roof cover, filled between with concrete. The middle and bottom drawings are examples of semifireproof systems, in which tile protects a timber structure. The middle one shows a tile barrier attached to the lower flanges of iron beams, protecting a plank floor. The bottom image shows a similar system, but made with hollow tile blocks. *The Inland Architect and News Record* 19 (June 1892).

plaster; the other method placed a cast iron column inside a metal tube, and filled the space between with plaster. The first method had the advantage of being comparatively cheap but was relatively ineffective.[92] The latter system was impractical as well as expensive, yet until the 1870s, columns made this way were the only kinds considered truly fireproof. Some authorities recommended them, and New York City's 1871 building law required such double columns in all cases where a column supported a wall (e.g., in a cellar, where the cellar extended past the building's facade, under the sidewalk).[93]

Architects Peter Wight and William Drake introduced the first practical system of column fireproofing. In 1874, they patented a system that consisted of a cast iron column in a cruciform ($\times$) shape with wooden wedges filling in between the flanges. How well the wooden wedges would have insulated the column in an actual fire was never to be discovered because for his order for the columns, Wight decided to make the wedges out of Loring's porous terra cotta. The wedges were held in place with small iron plates that screwed into the flanges of the columns and fit into indentations in the corners of the tiles. Wight finished the column with a coat of plaster, topped it with a terra cotta capital, and then painted it.[94] The first building to have these columns, the Chicago Club House (completed in 1876), was relatively small, but its owners went to the expense of making it partly noncombustible because they had lost two earlier buildings to fire, one in the 1871 conflagration and the next in 1874.[95]

Although cruciform-shaped columns were not as strong as round, hollow columns, Wight chose this obsolete shape to avoid the problem of attaching fireproofing material to a round column. Around 1879, Wight designed a round, hollow column with small projecting flanges (about one-and-a-half inches) cast on its walls, which he used to hold the tile plates. Wight also made tile wedges to fit Phoenix Iron Company's round, wrought iron columns, which already had flanges. At the end of the 1870s, New York City's building department accepted his fireproof column as a substitute for the double column. By the early 1880s, Wight had installed his columns in half a dozen buildings in Chicago, buildings in several Midwest cities, and the Patent Office in Washington.[96] Wight also introduced a system for fireproofing round columns, which involved semicircular tile plates. Invented to meet the requirements of the federal government, the curved, one-and-a-half-inch-thick plates were held in place with iron bands that fit into grooves along the upper and lower edges of the plates. This type of column went into eighteen federal buildings constructed in Western states between 1880 and 1886.[97] Other manufacturers offered curved tiles for fireproofing round columns, but often they did not secure them and relied simply on friction and a coat of plaster to hold the tile. Such tiles could be more easily dislodged in a fire than Wight's products; nevertheless, the cheaper systems eventually drove out the older ones[98] (Fig. 3-6).

A novel application of the porous tile was in a product called "terra cotta lumber." Charles C. Gilman of Iowa patented this idea in 1881; his patent described slabs made of kaolin clay and sawdust that burned out in the firing. Unlike hard or even most other porous terra cotta, the resulting fine-grained clay boards could be sawed, carved, drilled, and nailed, and thus could be used in place of lumber for floor joists, partition walls, and

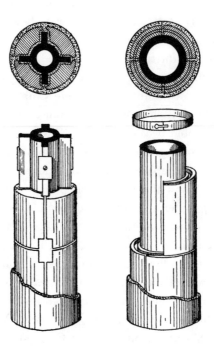

Fig. 3-6. Tile column fireproofing from the 1880s. *Left*, a hollow cast iron column with flanges. Small rectangular plates screwed into the flanges held the tile fireproofing wedges. *Right*, curving tile blocks designed for round columns. Iron bands that fit into grooves in the top and bottom edges of the tile held the tiles in place. The columns would be finished with a layer of plaster or cement. J. K. Freitag, *Fireproofing of Steel Buildings*.

roof sheathing as well as for column fireproofing. It also resisted heat and muffled sound better than solid tile. The first company to manufacture Gilman's product, the New York Terra Cotta Lumber Company, began operations at Perth Amboy, New Jersey, in 1882. Gilman licensed his patents to the company and became its president. The company also marketed its license to the product, along with shop space and land for entrepreneurs. Terra cotta lumber did not catch on immediately. According to a U.S. government report, the original version of terra cotta lumber—made of a fine clay, which could only be ground with high-powered machines and had to be fired at very high temperatures—cost a great deal to make. Soon a Minnesota company introduced a variation in the manufacturing process that utilized cheaper surface clay. Eventually, other companies manufactured this product and by the 1890s, terra cotta lumber had become a commercial success.[99]

The tile industry passed another milestone in the early 1880s when the tile manufacturer Henry Maurer made hollow arch floor blocks out of porous terra cotta, for the attic of the south wing of the Patent Office in Washington.[100] This was the first time the entire load-bearing tile span was made of porous tile, although the skewback blocks in the floor of the Mitchell Building in Milwaukee (c. 1878) had been made of this material. By the 1890s, porous tile became a standard material for tile arch blocks.

The inventors, manufacturers, and architects who founded the American hollow tile industry all had worked on iron and brick fireproof buildings and knew one another. As previously noted, Johnson and Kreischer jointly patented a tile block. Johnson patented a roof tile with William J. Fryer and also assigned to Fryer one half of his tube-and-binder floor patent. Heuvelman and Haven worked together at an architectural iron firm that supplied iron for fireproof buildings before teaming up and branching out into the hollow tile business. Their first hollow tile contract was for the New York City post office, where Fryer's company, Aetna Iron Works, had a contract to supply iron.[101] The investors and directors of fireproofing companies often had interests in several companies that made materials for fireproof construction. For example, Jose de Navarro, who became a Portland cement manufacturer, served as a director of the New York Terra Cotta Lumber Company. Navarro used terra cotta lumber in the huge, fireproof apartment complex in New York City he developed, known as the Spanish Flats (1882–85). Another director of Terra Cotta Lumber Company was Adolphus Bonzano, the chief engineer with the Phoenix Bridge Company. In addition to designing and manufacturing bridges, Phoenix

Bridge Company made iron roofs for fireproof buildings, built with structural shapes rolled at the mills of the Phoenix Iron Company.[102] It was the men who worked on fireproof buildings—the designers and iron manufacturers—who advanced fireproof construction technology, by creating new products, improving existing ones, and investing in each other's companies.

While hollow tile made modest headway in the 1870s, manufacturers introduced a less costly system of fire-resisting construction that gained wider market acceptance. In this "semifireproof" system, fire-resistive materials protected a building's ordinary wooden frame, which made it safer, although not fireproof. The idea of protecting wood with noncombustible coverings was an old one, and builders continued to use the simple strategy of putting a layer of mortar on wood floors for fire protection. In an attempt to improve on this, R. M. Hoe, the well-known manufacturer of revolving cylinder printing presses, devised a semifireproof floor that employed iron and plaster cladding. His 1868 patent called for fastening iron sheets to the bottom of joists and filling the space between the joists with plaster; the joists were covered on top with iron sheets, then a layer of plaster and a wooden finish floor.[103] Hoe installed this floor in the new plant he built in New York City around this time; whether anyone else adopted his idea is unknown.

The semifireproof systems introduced in the 1870s were more practical than this and employed the new materials of the decade: terra cotta, concrete, and metal lath and plaster. As designers came to understand that all structural materials, whether combustible or noncombustible, had to be insulated from fire, some reasoned they might just as well protect wood as expensive iron members.[104] While some people called buildings with protected wood frames "fireproof," departing from how the term was used, more commonly they were described as "semifireproof," "partially fireproof," "second-class fireproof," and even, incorrectly, as "slow-burning"— an acknowledgment that they were different from, and less secure than, thoroughly noncombustible buildings. An advantage of the semifireproof systems was that they could be installed in existing buildings, not only new construction. The term *fireproofing*, meaning something *applied* to protect structural members, entered the builders' vocabulary at this time.

One of the fireproofing systems that became popular in Chicago in the 1870s and 1880s was made of metal lath covered with concrete or plaster. Metal lath had been used for many years as a fire-resistant substitute for wood lath. New kinds of lath came on the market in the second half of the nineteenth century, including wire lath, also called wire cloth, and later, expanded metal, made by slitting iron sheets and then pulling each side of the sheet outward to create a metal mesh.[105] Around 1874, James John, a Chicago builder, began installing a fireproofing product that became his specialty—galvanized wire and concrete, with which he made interior partitions and ceilings.[106] The system continued to be popular in Chicago through the mid-1880s. Buildings on which he worked included the Continental Building and the Cooper & Carson Building, both under construction in 1884. Although he did not patent this product, John did patent another fireproofing idea: hollow plaster blocks that fit in between wood joists and made an airtight floor.[107]

But the most widely used system for fireproofing wood framing was tile plates. In an early form, in the 1870s, the plates were thin, perhaps one-and-a-half or two inches thick, and hollow.[108] In the 1880s, solid pieces of terra cotta lumber were used for this purpose. Inventors created many variations of this system, and the Patent Office granted numerous patents for fireproofing tiles in the 1870s. One of these patents, granted to George Johnson in 1872, called for a flat, hollow block that slid over wood strips along the bottom of joists and filled the space between the joists.[109] The following year, the production superintendent with the Fire-proof Building Company, Lucien Tartiere, patented a similar system consisting of hollow blocks with angled sides designed to hang from triangular holders along the joists.[110] Tartiere's tile represented an improvement over earlier systems because it covered up, and thus protected, the bottom of the joist. Henry Maurer, a fire brick manufacturer like his uncle Balthasar Kreischer, patented a tile block that resembled Johnson's except that it had internal webs, which made it stronger.[111] Peter Wight patented a fireproofing ceiling-tile that was held in place only at the corners, which made it possible to remove a single tile without disturbing the others.[112] The tile systems differed from common ceiling coverings such as tin plate in that they created airtight surfaces and were capable of staying in place when water at high pressure was thrown on them (Fig. 3-5).

A New Fireproofing Regime
in the 1880s

In the upswing in construction activity that coincided with the business recovery in the early 1880s, owners chose to use the new, hollow tile systems—either fireproof or semifireproof— more often than brick arches when they wanted a fire-resistive building, especially for important structures. Owners of a greater range of structures—office buildings, warehouses, hotels, apartment houses, and, notably, theaters—elected to build with one of the new fire-resistive systems. In New York City, according to the architect and architecture professor Theodore M. Clark, "burnt clay" was "universally used in the shape of blocks, made hollow, to save weight, and tapered so as to form a flat arch. . . . The old-fashioned brick arches, turned upon the lower flanges of iron beams, have now entirely disappeared from ordinary work, except for carrying sidewalks over coal vaults."[113] In 1885, Peter Wight noted that since the introduction of "improved methods of fireproofing," Chicago had added eighteen fireproof buildings to its stock. Half of these were under construction in 1885; of the nine newest buildings, six were iron and tile and three were wood frame and tile.[114]

The new fireproofing systems spread outside Chicago and New York, home bases of the hollow tile industry. Philadelphia's pioneer modern office block, the Wood Building (c. 1881), contained hollow tile floor arches from Henry Maurer & Son.[115] In Washington, several projects in the early 1880s contained hollow tile: the reconstruction of the third floor of the south wing of the Patent Office, reconstruction of sections of the Smithsonian Castle, and the Pension Building (1882–87). Adolf Cluss, an architect who had worked as a draftsman in the Construction Bureau of the Treasury Department during the Young/Bowman era, handled the first two of these projects. At the Patent Office, he removed the attic floor and roof of Mills's south wing and rebuilt them with hollow tile manufactured by Henry Maurer's company. Cluss used hollow tile to make part of the deck of the new west wing roof at the Smithsonian Castle.[116] Wight Fire-proofing Company and Pioneer Construction Company won contracts to put their tiles in many of the federal buildings that went up in Midwest in the 1880s. Customers around the country also bought the semifireproof systems. Wight Fire-proofing Company featured ceiling tile, rather than arch blocks, in its advertising and installed this product in buildings in Chicago, Minneapolis, and New York. The Pioneer Company also did semifireproof work in cities around the Midwest.[117]

With their customer base growing, the "fireproof building" industry expanded. One of the three pioneer hollow tile firms, HHC, left the business by the 1880s, but several new ones entered the ranks. By 1884, the Pioneer Fireproof Construction Company, successor to George Johnson's company, had established a substantial operation in Ottawa, Illinois, with water- and steam-powered clay crushers, material movers, and tile presses. The Fire-proof Building Company of New York enlarged their works.

Henry Maurer started a fireproofing company in 1875, after patenting a hollow tile for semifireproof construction, with works in Perth Amboy, New Jersey, an area rich in clay. The Raritan Hollow and Porous Brick Company set up business in this part of New Jersey in 1882. Raritan made hollow floor and partition blocks, as well as terra cotta lumber. Peter Wight founded the Wight Fire-proofing Company in about 1881 and opened offices in Chicago and New York City. Wight probably did not have his own manufacturing facilities, and therefore contracted with tile manufacturers to make products to his specification. Some companies also made fireproofing products out of concrete. One of these, the New York firm of John J. Schillinger, offered concrete blocks for fireproof floor arches, partitions, and furring, and concrete plates for fireproofing wooden beams. Beginning in 1879, Schillinger took out patents for a number of fireproofing products. According to a contemporary, concrete blocks for fireproofing "are made in most of our large cities."[118]

Another important item introduced in this period, designed to protect iron beams, was a covering for the bottom flanges. Because of wrought iron's unwarranted reputation for being less liable to be damaged in a fire than cast iron, designers typically left the lower flanges of wrought iron beams unprotected, even after they started to put insulating coverings on cast iron columns.[119] A method for protecting beam flanges, patented in 1872, involved a new design for the bottoms of skewback blocks so that they extended underneath the bottom of a beam. The idea was first put into effect in the Mitchell Building in Milwaukee, around 1878. A few years later, Henry Maurer produced a style of hollow tile floor arch with flange-protecting skewback blocks. Maurer's arch continued to be in demand at the end of the century, even though simpler (and therefore less expensive) methods of covering beams were available by then.[120] Other companies also offered hollow tile arch styles with beam-protecting skewbacks. These blocks had the disadvantage that the small projecting piece was liable to break.

In 1883, Peter Wight patented a simpler solution to the problem: a separate piece of tile that covered the flange. The distinctive feature of Wight's tile was that it extended up and covered the sides of the flange; skewback blocks on either side of the beam held this "soffit tile," as it was called, securely in place. Wight first used soffit tiles in a new building for the Mutual Life Insurance Company in New York City (1883–84) and afterward put them in all but one of the floor arch contracts executed by his company. Other tilemakers in the West turned out their own versions, but

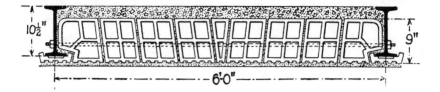

Fig. 3-7. Separate soffit tile placed under the lower flange of the I-beam to insulate the beam from fire. First used in the floor arches of the Mutual Life Insurance Building, New York City (1883–84), illustrated here. J. K. Freitag, *Fireproofing of Steel Buildings.*

the typical commercial soffit tile covered only the lower beam flange, not the sides[121] (Fig. 3-7).

A final development in the fireproofing business in this period was that bridge shops began to bid on contracts for the structural iron in buildings. Before the 1870s, contracts for iron building frames usually went to architectural ironworks. After the Civil War, bridge shops—manufacturers of iron bridges—entered the world of iron buildings, initially by bidding on contracts for iron roofs. Phoenix Iron Company made iron roofs in the 1860s, including the new 1866 roof for the Smithsonian Castle in Washington. In the early 1870s, the Kellogg Bridge Works of Buffalo won the roof contract for New York City's post office.[122] At the beginning of the 1870s, building laws in Boston and New York required roofs, in certain circumstances, to be constructed in a fireproof manner which, as a practical matter, meant framed in iron.[123] Soon, bridge companies bid on ironwork for the other parts of a building—the columns as well as beams—which up to this time had been the domain of foundries.

Before the great Chicago fire, a fireproof building was simply a noncombustible building. After the fire, experts refined the definition: to be considered fireproof, a building's constructive materials had to be nonconducting as well as noncombustible. Peter Wight summarized the new understanding in 1879 when he explained that despite what was once supposed and what some architects continued to believe, "fire-proofing depends more upon the *protection* of the materials of construction than upon the materials themselves."[124]

By the mid-1880s, at least half a dozen American companies manufactured hollow tile for fireproof construction and offered a variety of floor blocks: heavy, light, with or without interior webs, for flat or curved arches.

Although by no means a commodity yet, hollow tile blocks were on their way to becoming one. At the same time, the number of rolling mills that could make solid beams had increased, and on the East Coast, imported beams could be had at competitive prices. Bridge shops became a source for ironwork in buildings, and eventually displaced the architectural iron foundries.

Despite all these developments, fireproof buildings still constituted only a small share of all new buildings, even in New York City. In Boston in 1894, according to a fire insurance executive, iron and tile buildings were "so rare that each one is looked upon by the public as an expensive curiosity."[125] The innovations of the 1870s and 1880s improved the materials of fireproof construction, but did not close the price gap. What was an owner interested in greater security from fire, but reluctant to go to the expense of erecting a fireproof building, to do? He could use a semifireproof system. Alternatively, he could adopt the system developed for New England textile mills, which its advocates believed, "will almost rival the brick-and-iron floor [building] in its fire resisting qualities."[126] In the 1880s and 1890s, this system, known as "slow-burning construction," rose to prominence as an alternative to iron and tile.

4

MILL FIRE PROTECTION METHODS ENTER THE MAINSTREAM

There is practically no such thing as a fire-proof building. . . . We have reached the conviction that the abundance of wood in the United States may not only be made to serve economy in construction, but may be made as safe as the so-called fire-proof buildings of Europe, if architects and builders are rightly instructed in the use of the material.

EDWARD ATKINSON, fire protection expert, 1880

British manufacturers in the late eighteenth century developed the iron and brick arch system to defend against the fire hazards posed by mechanized textile production. American businessmen, trying to become successful textile manufacturers themselves, eagerly borrowed and adapted British methods; but one measure they left alone was fireproof construction for factories. With all the difficulties and uncertainties in starting a new business, investors had no desire to overspend on buildings. Indeed, most British investors felt the same way: until about the middle of the nineteenth century, only a minority of manufacturers built fireproof textile mills and warehouses. The typical mill owner might put up a fireproof room or a small fireproof building to house the most hazardous operations or to store valuable stock. But the vast majority of British mills in the first half of the nineteenth century had conventional timber framing inside.[1] In American textile factories, even fireproof rooms were rare. Indeed, few firms in any industry adopted iron and brick construction.

Yet American mill owners did not ignore the problem of fire. Rather, over time they developed a fire-resistive method of construction that avoided using noncombustible materials. The main principles of the system were strict compartmentation—separating horizontal from vertical space—and on-site fire-extinguishing capabilities. They built their timber-framed mills in such a way that they would burn slowly in a fire, and, with fire-fighting equipment on hand, operated by the mill's own workforce, fires could be put out quickly. In short, mill owners created a fire-safe system that cost much less than fireproof construction. The system became the standard one for building textile factories and eventually spread to commercial architecture, as a fire-resistive alternative to "absolutely" fireproof construction.

Called "mill construction" or, more descriptively, "slow-burning construction," the system originated in New England, the nation's largest textile-producing region. Its creators considered it a balance among a mill owner's various requirements, or, as a contemporary engineer described it, "the best fitness of means to ends in the effort to reduce the cost of production to its lowest terms."[2] In the first part of the nineteenth century, mill design evolved unsystematically, with builders and owners trying out new or borrowed ideas. But in the last quarter of the century, roughly, an ideal, or "standard mill," emerged. Its features were formulated by a unique institution: an association of mutual fire insurance companies that covered factories exclusively, known collectively after 1888 as the Associated Factory Mutual Fire Insurance Companies (AFM). The "mill mutuals" set minimum standards and only insured properties that met their fire-safety criteria. But their low rates, compared with those offered by mainstream (the investor-owned or "stock") fire insurance companies before the 1890s, prompted many manufacturers to comply in order to save on fire insurance.[3]

Around 1880, the ingredients of the standard mill—which included masonry walls, heavy timber framing, compartmentation of the interior, and provision of fire-extinguishing equipment on-site—were codified and named "slow-burning construction." AFM underwriters argued that mills built to their standard were as safe as English iron and brick fireproof mills, a claim bolstered by their very low insurance rates. Moreover, slow-burning construction methods could be applied to other kinds of buildings. Indeed, AFM engineers and officials considered slow-burning construction superior to fireproof construction because it accomplished the same goal at a lower cost. As the civil engineer and AFM inspector, John R. Freeman, observed, "it is not much worse to burn up a dollar than to bury it beyond circulation in an over-expensive incombustible structure."[4] The popularity of semifireproof methods among commercial property owners suggests that many were willing to invest in safeguards against fire, but they balked at the cost of noncombustible construction. Such owners constituted a pool of potential clients for slow-burning construction. However, they first had to find out about it, since those who had no connection with textile manufacturing would never have come across it. Similarly, architects rarely designed factories and also would not have known about mill construction methods.

To one man, Edward Atkinson, goes much of the credit for bringing slow-burning construction to the attention of city landlords and architects, and indeed to the reading public. For twenty-seven years, Atkinson served as president of the largest of the mill mutual companies. Through his company's publications, his articles in national magazines, and his speeches, Atkinson convinced many Americans to accept slow-burning construction as an effective means. Moreover, the organization he helped establish and imbued with his waste-not ethic significantly influenced safety practices in the United States for the better.[5]

Origin of the Factory Mutual Fire Insurance
Companies and Slow-burning Construction

While it may seem logical that the fire insurance industry would be a force for improving the fire safety of buildings, this was not the case. To a stock fire insurance underwriter, it made no difference how many individual properties burned so long as he earned enough in premiums to cover losses and expenses, and pay dividends to stockholders. In fact, the risky properties for which he charged high premiums, could be profitable for agents and companies: they brought the former higher commissions and the latter more income to invest. Stock companies had no business reason to try to get owners to alter their properties to make them safer; rather, underwriters simply charged owners according to the condition of properties as they were. They set rates in a rough and ready fashion, grouping properties by gross characteristics, such as the materials of construction (masonry or frame) and use (occupancy); they rarely made distinctions among the properties in the broad categories. As long as the rates they set for various classes produced enough income overall, the variations among the properties in a class were immaterial. In certain hazardous types of occupancies, such as textile mills, companies might charge an owner more if they found conditions that increased the chances of fire and loss; but they rarely reduced rates for owners who introduced safety measures—who made their properties better than average.[6] This situation discouraged owners from investing in safety: they got no break in their insurance costs while the insurance companies reduced their own exposure. Moreover, stock underwriters did not encourage owners to make improvements.

Mutual fire insurance companies, in contrast, were owned by policy-holders rather than stockholders, and therefore operated according to a different set of incentives. Where stock companies sought to bring in as much revenue as possible, which they invested until it was needed to pay expenses, mutuals endeavored to reduce the revenue they needed, and thus the cost of insurance for members, by keeping losses down. Stock companies spread their risk by insuring a large and diverse pool of properties. Mutuals could reduce their exposure by being selective and covering only properties unlikely to burn. An early mutual company, and also the nation's first successful fire insurance company, The Philadelphia Contributionship for the Insurance of Houses from Loss by Fire, refused to cover buildings it considered hazardous such as wooden buildings and theaters, and famously, any house with a tree growing nearby. The no-tree rule, eventually abandoned in 1810, arose from the concern that trees would obstruct the work of firefighters and could spread a fire in the street to the house; it led to the felling of many trees.[7] But instead of insuring low-risk types of properties, like homes, the mill mutuals insured high-risk properties—mills and associated buildings—although only the *low-risk* mills.[8] The mill mutuals were the first insurance companies to specialize in insuring industrial properties.

While textile mills—with their dangerous combination of combustible stock and numerous sources of ignition—were among the most hazardous kinds of factories, nevertheless fires could be prevented and destruction limited. Owners could reduce danger through good management (e.g., by properly storing and removing combustible waste, and isolating the most fire-prone operations). A substantially constructed mill, watchmen, and a supply of water for fire fighting would help stem losses when a fire did break out. Yet fire insurance companies took little notice of safety measures in a mill when setting a rate for it. This lack of discrimination exasperated the inventor and philanthropist Zachariah Allen: around 1834–35, a period of rising insurance rates for mills, Allen tried to bargain for a rate reduction on his Rhode Island woolen mill, where he had installed a novel system of a water pump, pipes, and hoses for fire fighting. But he could not convince any underwriter to visit his mill and evaluate his improvements. Angered by this rebuff, Allen then worked to set up a mutual fire insurance company that would insure high-quality mills like his own.[9] Allen did not originate the idea of a mutual company for factories. He wrote that "there had been several mutual insurance companies established by manufacturers of New England" before 1835, although all had failed.[10] One of

the companies, which Allen described as "one of the last" of the early factory mutuals, turned over its business to Allen's new company. This was
probably the Worcester-based Manufacturers Mutual Fire Insurance Company (MMFIC), which closed in 1836 after only two years in operation.[11]
Allen's Providence, Rhode Island–based firm took the same name. Considering the poor track record of its predecessors, the wonder is that the new
MMFIC survived. Its requirements for membership were modest and its
rates were comparatively low. Allen did put in place a "more vigilant system" of inspection than was customary in the insurance industry. Nevertheless, growth was slow and for thirteen years, MMFIC had the field of
mutual fire insurance for factories to itself.

Then, in 1848, during another period of great cost pressures for textile
manufacturers, a second factory mutual started up in Rhode Island.[12] Allen
viewed the new company as a collaborator rather than a competitor. In
these days before reinsurance, fire insurance companies typically limited
the amounts they would write on any single property, to avoid being ruined by one bad fire.[13] For this reason, several insurance companies covered
valuable properties, each one writing some portion of the insurance the
property owner wanted. Probably most of MMFIC's customers had insurance from several companies, and MMFIC had to place the amount that
exceeded its limits with stock fire insurance companies. If the number of
mill mutual companies increased, MMFIC could place some of this excess
with mutual companies, at better rates. A proposal by a group of Boston
businessmen to establish a factory mutual company in Massachusetts
tested Allen's commitment to seeing companies multiply, rather than expanding MMFIC and raising its coverage limit. But he held firm. It was
"far better to enlist a double number of efficient inspectors of the mills,
and to render available the personal knowledge of the risks, which a
greater number of directors will combine." He believed the shared interests of the companies "would necessarily establish a communication of in-
telligence . . . such as now exists between the two mutual offices" in Providence, "the common object being a reduction of insurance premiums to
the actual cost on the best cotton mill establishments in New England."
Their "mutual surveillance to check carelessness for the common benefit"
should result in "a very excellent system of mutual insurance."[14] Chartered in 1850, the Boston company, Boston Manufacturers Mutual Fire Insurance Company (BMMFIC), became the largest of the AFM companies
in the nineteenth century. Four more Massachusetts and Rhode Island-
based companies—Arkwright, Firemen's, State, and Worcester—started

up between 1854 and 1860. All served New England customers almost exclusively, mainly textile manufacturers. While each company had its own charter, governing board, and rules, they did indeed become collaborators rather than competitors as Allen expected they would. All insured only well-built and well-managed mills.

The AFM companies won customers by charging lower rates than the stock fire insurance companies. From the 1830s through the 1850s, stock companies' rates on mills never fell below 1¼ percent per $100 of insurance; 2 percent was typical, but could rise to as much as 4½ percent, and this for fire coverage alone. In the early days, AFM companies set their rates at 75 percent of that charged by one of the larger stock companies, the Aetna Insurance Company of Hartford, with deductions from and additions to the price depending on the characteristics of a building. The final cost of insurance usually would be less than this, because the mutuals returned excess revenue collected to the members. In most years, members received rebates and these, coupled with the already reduced-price premiums, made their actual insurance costs very low indeed.[15]

This saving came with obligations for AFM members. Besides having to maintain their mills to the companies' standards, mutual members had to pledge to make up any annual financial shortfall, should losses and expenses exceed premium income. For members to assume this sort of risk—one which in stock companies fell on the investors—they had to have confidence in the mutuals: to trust that the companies would inspect risks carefully, enforce standards uniformly, set premiums properly, adjust losses fairly, and be thrifty. By the same token, mutual officers trusted members to make good on their "deposit notes" or pledges, in case expenses exceeded income.[16] This trust developed from personal ties among the directors, executives, staff, and members, who were business associates, neighbors, and relatives of one another. And instead of finding it an unwanted imposition, many members welcomed the visits of the mutuals' inspectors, who were knowledgeable mill-men, and the safety advice they offered.

The companies went their separate though similar ways, writing policies on the same properties and sharing a business philosophy and information, but without having uniform procedures. By the 1870s, the number of AFM companies had increased to seventeen.[17] AFM members paid much less for insurance than did other mill owners; for example, in its first twenty-seven years of existence, the BMMFIC returned two-thirds of premiums collected to members.[18] Also in this decade, the destruction of several valuable mills by fire prompted the companies to work together more closely. Officers of several companies began to meet regularly to adjust losses, issue joint circulars, and discuss underwriting standards. The year 1878, when BMMFIC elected Edward Atkinson as its president, proved to be a pivotal one for the AFM companies. Under his leadership, the AFM organization gained national attention for its efforts to reduce fire loss.

Edward Atkinson—pacifist, feminist, economist, and indefatigable publicist for all he considered good—shaped the mutuals into a unified organization and a force that changed the fire insurance industry in the United States. His apparently diverse causes shared a common theme: an abhorrence of waste and inefficiency. Fire insurance may seem like an unpromising field for a visionary, but he saw ways to reform the industry, just as he found ways to improve so many areas of human endeavor. Instead of being a means to compensate for loss, Atkinson defined fire insurance as a means to reduce "fire waste"—that huge amount of property unnecessarily lost to fire every year. Given proper guidance and incentives, Americans could stem this loss, and Atkinson spared no effort to provide both. A memoir of Atkinson in an insurance journal likened him to the type of "New England old-time clergyman" portrayed by Harriet Beecher Stowe in her novel, *The Minister's Wooing*. Such a man "felt irresistibly bound to 'testify' as soon as he saw the light upon the moral side of a subject. Mrs. Stowe likened her personation to an 'honest old granite boulder,' impelled to roll with all its might toward any wrong thing, regardless of the consequences. Mr. Atkinson had a similarly uncompromising sturdiness of conviction."[19] As this characterization suggests, Atkinson's self-assurance invited scorn and ridicule, yet his critics often resorted to ad hominem attacks because his arguments were so persuasive.

In the early years, although the AFM companies tried to accept only the best mills, they could not be too picky. They mainly required that factories be clean, and have watchmen and a source of water for fire fighting. At this time, it was the mills in Lowell, Massachusetts, the nation's foremost textile city, that served as models of factory fire safety. Lowell's mill owners had such confidence in their fire protection measures that at first they carried no fire insurance at all. In 1850, following several mill fires, they set up their own mutual company—chartered, coincidentally, in the same month as BMMFIC—and hired the civil engineer James B. Francis, who also managed waterpower as chief engineer with the Proprietors of the Locks and Canals company, to oversee fire protection.[20] Francis's pioneering work to reduce fire loss for Lowell's mill owners, which included studying the causes of fires in order to prevent and control them, shaped the philosophy and practices of the AFM companies.

Lowell's Proprietors put conditions in the terms of sale of mill sites to assure that new mills met certain construction standards.[21] The earliest Lowell mills, built in 1823, contained standpipes—vertical water pipes with outlets at each floor to attach fire-hoses—supplied by water tanks in the attics. Waterwheel-powered force pumps, which pumped up water from a nearby river, canal, or reservoir, kept the tanks full. After a destructive mill fire in 1828, Lowell's mill owners put hydrants in their yards, also connected to the roof tanks. Two decades later, to improve water pressure and to bring more water to a burning building, the owners linked the pipes in the separate mill yards and built a reservoir to increase the water supply to the network. This reservoir and pipe system predated the city of Lowell's own municipal waterworks.[22]

Other fire protection measures at Lowell's mills included sprinklers, tower stairways, and exterior ladders. The first sort of sprinklers consisted of horizontal pipes with holes drilled in them that ran along the ceilings and were fed by rising pipes attached to the water main in the mill yard. They operated when someone opened a valve, which would start a pump that pushed water up the riser and into the perforated pipe. This idea had been borrowed from England, where a patent for pipe sprinklers was granted in the early nineteenth century.[23] In 1845, the Suffolk Manufacturing Company installed such a sprinkler in its picker room, the first Lowell mill to have one. By the early 1850s, many of Lowell's mills had sprinkler pipes in their attics, and before the end of the decade, owners had installed them in carding, picking, and spinning rooms, and inaccessible spaces. Stair-towers, occasionally called "porches" by contemporaries, appeared on some of the earliest American mills, such as the second Boston Manufacturing Company mill at Waltham, built 1816–19.[24] Putting the stairway outside the building made more floor area available for productive operations inside and allowed the stair-tower to be isolated from the floors by fire-resistive doors. This eliminated an avenue for fires to spread upward, and in addition, created a protected place from which firefighters could fight a fire. Mill designers eventually installed standpipes and hoses in the towers. Ladders and platforms were commonly found on the walls of New England factories, including those at Lowell, in the first half of the nineteenth century. Although they look like fire escapes, their purpose was to enable firefighters to get to a fire from outside a building. Such fire protection measures were peculiar to America. A British millman, James Montgomery, who visited the United States in the late 1830s to study American textile manufacturing, regarded these fire protection measures as features that distinguished American from British practice.[25]

Another characteristic of Lowell mills was plank floors, made of heavy boards laid directly on girders. This joist-less style of floor created a smooth ceiling underneath that did not require plastering. They could be found in mills throughout New England by the end of the 1830s, and Montgomery included plank floors among the distinguishing features of American mills. He apparently was unaware that the idea originated in England.[26] The idea that wooden floors could be fire-resistive emerged from the observation that large-dimension timber tended to char to a certain depth in a fire, at which point the charcoal slowed combustion, allowing the member to stay in place longer.

Although these elements—sprinklers, stair-towers, exterior ladders, plank floors—came to be widely adopted in New England, no one considered them parts of a system. The first stab at codifying the characteristics of a fire-safe mill occurred at the end of the 1850s, when BMMFIC's secretary put into writing his notion of a first-class cotton factory—the kind that would be charged the lowest rates. Such a mill would have stone or brick walls; a roof covered with slate, metal, or shingles laid in cement; standpipes and water pails; elevators walled off from the floors; fixed ladders and platforms; steam heat; and sprinklers "wherever deemed to be necessary." The company wrote policies on mills that departed from these guidelines, but after 1856, it refused those with wooden walls.[27] In 1858, BMMFIC added several more items to the list, including plank floors, which it advised owners to deafen with plaster and to plaster underneath.[28]

In the early 1880s, Atkinson compiled all the then-current recommendations to create a definition of an ideal mill, which he called a "standard" mill. BMMFIC henceforth only insured mills that conformed to the standard. Atkinson publicized the standard mill in reports issued by BMMFIC; the earliest descriptions were in Special Report No. 5, dated September 1881, and in No. 10, dated April 1882. He gave the system its evocative name, slow-burning construction. Also at this time, the first American textbook on fire safety for factories appeared, *The Fire Protection of Mills*. Written by Charles J. H. Woodbury, BMMFIC's college-educated engineer/inspector, it outlined the ingredients of a slow-burning mill.[29]

The standard evolved over the following few decades, as ideas about how to improve safety changed, but certain elements remained constant. These included brick walls formed into piers between the windows that supported the ends of the floor beams: large timbers, usually sixteen to twenty-five feet long, set eight to ten feet between centers. Inside the building, the beams rested on cast iron caps atop wooden posts. The most characteristic feature of the slow-burning mill—the floors—were made of planks three to four inches thick, with their sides grooved and filled with wooden splines, and covered with a finish floor of one-and-a-quarter-inch-thick boards. The splines prevented gaps from developing as the plank dried and shrank over time. The building's roof was nearly flat and made much like the floors, with two-and-a-half- to three-inch-thick plank, and was covered with gravel, duck cloth, or tin roofing material (Fig. 4-1). The practice of separating vertical spaces, such as stairways and shafts for belts that drove the machinery, from the main floors, was carried out more thoroughly in slow-burning mills than in fireproof buildings, where designers mistakenly relied on the noncombustibility of construction materials and neglected compartmentation. In slow-burning mills, hatches and doorways were fitted with fire-resistive covers. A standard mill could not contain any hollow spaces, such as those above ceilings and behind plaster finish walls found in ordinary buildings, where fires might spread unseen and beyond the reach of water.

Since wood structural members, however large, eventually would succumb in a fierce blaze, fire suppression was an essential part of the system. Fire-extinguishing measures included buckets of water distributed around a factory; a source of water, such as a pond, river, or reservoir; standpipes with hose outlets for fire fighting; and internal sprinklers. Many factories also had private fire brigades composed of mill-hands, and every operative was expected to grab a bucket and douse a fire in an emergency.[30] The joist-less mill floor, with its widely spaced girders, could be easily swept with streams from hoses and spray from sprinklers. In the early 1880s, Atkinson and Woodbury recommended the recently introduced "automatic" sprinklers over perforated pipes. In automatic sprinklers, the horizontal pipes were fitted with little outlets, or sprinkler heads, along their length, instead of simple holes. The heads were sealed with solder and therefore the pipes could be kept filled with water under pressure, ready to be released should the solder melt.

Modifications of the standard began to be based on the results of experiments as much as observation of the consequences of fires. On taking over as head of BMMFIC in 1878, Atkinson studied the detailed reports of mill fires prepared by the company's chief inspector, William B. Whiting, and for the first time compiled statistics on the locations and causes of fires. He

STANDARD MILL CONSTRUCTION.

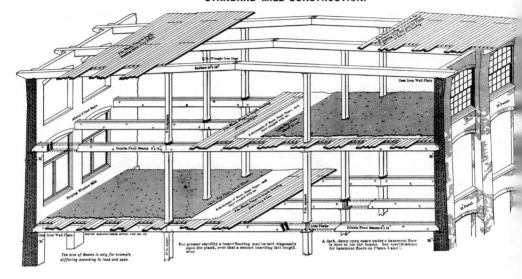

Fig. 4-1. A standard mill circa 1899, illustrating the details of slow-burning construction. The structural features of the system include brick walls formed in piers; timber posts; floor girders made of two beams; and plank flooring, laid directly on the girders. Not shown are the sprinklers, standpipes, roof water tank, fire alarm, and other measures that were integral parts of a standard mill's fire protection system. Boston Manufacturers Mutual Fire Insurance Company, "Standard Mill Construction," 11th ed., June 1, 1899.

found that many fires appeared to be caused by the lubricating oil used on mill machinery, and therefore commissioned a chemistry professor at the Massachusetts Institute of Technology (MIT), John Ordway, to find out why these machine oils burned so readily and what could be done about it. At the same time, Atkinson encouraged Charles Woodbury, a graduate of MIT's mechanical engineering course whom Atkinson hired to help with mill inspections, to conduct research. When BMMFIC moved to a new office in 1884, Woodbury set up a small area for experiments. Until the last decade of the nineteenth century, BMMFIC's laboratory was the nation's only ongoing fire prevention research operation.[31]

As knowledge increased, the AFM companies refined their standards and required old mills to be updated or lose their coverage. One of the most visible of these updates was changes in the roofs of older mills. By the last decade of the nineteenth century, the standard called for a nearly flat roof; owners of older factories removed their existing pitched and "double" roofs, built up the walls to the height of the peak at the gable ends, and put on a new flat roof. This had the added benefit of creating a usable top floor in place of a cramped and inaccessible attic. The double benefit was no coincidence: AFM officials always considered the economic and practical consequences of changes before recommending them. They gave each of the objectives of "safety, convenience, and stability" due weight.[32]

AFM's recommended construction methods were little known outside New England at the start of the 1880s. Few slow-burning mills could be found in the Philadelphia area, an important center of textile production, or west of the Hudson. Moreover, the system had not spread to nonindustrial buildings, even in New England, because people outside the mill world knew nothing about it. Edward Atkinson set about to change this situation.

As his memoirist made plain, Atkinson was not a man to withhold useful information. Since his company's standards had reduced fire loss for its members, why not for others? On taking over at BMMFIC, Atkinson began to issue circulars, pamphlets, and reports officially intended for members and potential members, but also aimed at a broader audience. In one early report, published in January 1880 and perhaps the first public statement of these specifications, Atkinson discussed his company's construction standards in the context of America's fire problem, which he suggested slow-burning construction could help solve. He also outlined an idea for safeguarding city centers that involved putting hydrants on the roofs of "high buildings" to help firefighters reach fires on the upper floors. Atkinson recommended standpipes for such buildings, something the mutuals required in factories but were rare in nonindustrial buildings. As importantly, Atkinson articulated the novel philosophy of the AFM companies, which was to prevent loss by fire.[33]

AFM technical staff helped Atkinson spread the gospel of fire prevention. Charles Woodbury and John Ripely Freeman, another MIT graduate Atkinson hired in 1886, were active in professional societies and gave papers about AFM's work and standards. In 1889, Freeman explained to a meeting of civil engineers that the business of the AFM companies was "as much for the purpose of preventing or reducing loss by fire, as it is for the purpose of adjusting or distributing the loss when a fire does occur."[34] They also taught the next generation of engineers and industrial leaders about mill construction and fire safety. In the 1886–87 school year, Woodbury spoke to engineering students in Professors Gaetano Lanza's and Peter Schwamb's class at MIT about AFM fire protection efforts. The following year, he delivered a lecture, "The Evolution of the Modern Mill," to students at Cornell University's mechanical engineering department, in which he discussed slow-burning construction. Freeman, who also lectured at MIT and Cornell, helped found a new field of engineering—what came to be called insurance, or fire protection, engineering.[35]

Slow-burning Construction
Spreads beyond the Factory

AFM officials were acquainted with iron and brick fireproof construction, not least because it was the principal one used to build British textile factories at the end of the century, but they dismissed it as an impractical solution. In his 1882 textbook, Woodbury remarked that "floors made of brick arches sprung between I-beams are so heavy and expensive as to be rarely feasible."[36] An overly costly mill tied up capital and constrained an owner who might want to rearrange a plant or even tear a mill down, if business conditions warranted this. What made slow-burning construction right for mills also suited it for many nonindustrial buildings. Atkinson argued that fireproof construction might do for "post-offices, custom-houses, or other government buildings, [and] . . . such structures as have been erected by life-insurance companies," which were built without regard to cost; but the great mass of ordinary buildings could be made safer without using noncombustible materials. A properly constructed wooden frame, he asserted, "may be made as safe as the so-called fire-proof buildings of Europe."[37]

Beginning in 1879, readers of the architecture magazine, *American Architect and Building News* (AABN), learned about slow-burning construction from editorials and readers' letters. That an architecture magazine should have devoted so much attention to this system—one used almost exclusively in New England and in textile mills, a kind of building that few architects ever designed—was due to a special conjunction of circumstances. First, the general advance in building technology at this time—the introduction of new material and structural systems—made many architects eager for technical information and AABN endeavored to provide this. Second, AABN's headquarters was in Boston, home of BMMFIC. BMMFIC subscribed to AABN, and Atkinson sent copies of BMMFIC's circulars and reports to the editors, who often commented on them. Atkinson in turn wrote letters in response to items he read in AABN, and the editors gave his views respectful attention. Through the pages of AABN, the principles of slow-burning construction received national exposure.

In 1879, AABN carried a long-running exchange between Atkinson and "C," an architect who never revealed his identity, which served as a platform for Atkinson to explain the slow-burning system. AABN's editor touched off the debate by complaining about the dreariness of New England factory towns, a condition he blamed partly on AFM underwriters because they discouraged mill owners from employing architects. He also acknowledged that from a fire safety standpoint, mills were superior to nonindustrial buildings, and therefore concluded, "unsatisfactory as the system is in its architectural results, it meets the demands for safe construction very successfully."[38]

In truth, mill underwriters took a dim view of architects, as Atkinson explained in a letter to the editor, because whenever they did design mills, they sacrificed safety and functionality for exterior effect. As proof that architects lacked the most elementary knowledge of structural fire protection, the buildings they usually designed—churches, hotels, schools, even the buildings in downtown Boston constructed after the 1872 fire—were, in Atkinson's opinion, almost without exception firetraps. Atkinson complained that AABN did little to educate its readers about fire safety: the latest issue, he noted, contained not a sentence about "the right construction of a building."[39]

In defense of his profession, architect "C" responded that contrary to what Atkinson believed, architects had both the knowledge and the desire to design safer buildings. The blame for poor construction, rather, lay with the owners, who refused to pay for better buildings. "C" estimated that a fireproof building would cost about 18 to 20 percent more than a common one and consequently, except for buildings with valuable contents, or the mills insured by the AFM companies, it was "next to impossible to induce owners to expend money in extra protection."[40] Atkinson replied to this by stressing that it was not necessary to spend a lot more to improve a building's safety. "Our mills are not fire-proof," he explained, "but if kept clean and in good order they are *slow-burning*. The contents may burn with great rapidity, but the structures themselves are built with a view to slow combustion."[41] Any owner could make his building safer by adopting the relatively inexpensive measures used in mills. Measures such as eliminating hollow spaces, and isolating shafts from the floors, could be implemented at no great expense. Moreover, public safety demanded them.

But since slow-burning methods had rarely been applied to nonindustrial buildings, no one could say exactly how much more they would cost. "C" believed the difference would be large and worried that clients would accuse architects who recommended the system of merely trying to boost their fees. "C" also made a point that architects and even insurance writers repeated endlessly: owners should be able to finance the higher cost of better construction through savings in their insurance premiums. "Why," he asked, "cannot the insurance companies establish a system of inspections, and make rates varying . . . according to the mode of construction?"[42] At the same time, he urged underwriters to support the passage of general building laws that would make safer construction compulsory.

Architects had other reservations about the slow-burning system apart from the cost. Were automatic sprinklers practical and reliable? How well would the system work in a commercial building, which did not have the kinds of backup protections—an independent source of water for fire fighting or a private fire brigade—that mills had? Atkinson considered these issues surmountable. He distributed copies of his circulars, building plans, and general specifications for slow-burning construction at no charge, as part of what he called his company's "missionary business."[43]

And in fact, architects started to use what they called the slow-burning system. "Mr. Edward Atkinson's agitation in favor of the cheap but 'slow-burning' construction . . . begins to bear fruit," AABN noted in November 1880.[44] As early as 1878—even before the publication of Atkinson's reports on mill construction or Woodbury's textbook—several Boston area buildings, including three warehouses, a bank, a school, several churches, and a house, incorporated slow-burning principles.[45] At the beginning of the 1880s, Boston architect William G. Preston completed one of the first large buildings designed to be slow-burning: the headquarters and exhibit hall of the Massachusetts Charitable Mechanics' Association in Boston. Preston had to persuade the owners to allow him to use the system, and he used it only "so far as the managers would consent . . . the system being less well-known then." Preston went on to apply it to several projects, including a shop for the Boston Terra-Cotta Company (1885), built "of mill construction throughout," and the Chadwick Lead Works (1887), both in downtown Boston; a library in the town of Lincoln; and a school for the handicapped in Waltham (1888–89).[46] Around 1885, the Massachusetts Institute of Technology erected a new structure in Boston "substantially" on slow-burning principles, and Cornell University in Ithaca, New York, also put up a slow-burning building. The research station at Woods Hole, Massachusetts, built a laboratory and boarding house according to specifications furnished by BMMFIC. Harvard Medical School commissioned a new building patterned on the "best type of American mill construction." Hospitals, too, adopted the system: Atkinson frequently recommended one in Waltham, Massachusetts, designed by his son William, as an example. Whenever he learned of a fire in a hospital, asylum, or college, Atkinson sent their administrators copies of BMMFIC pamphlets on slow-burning construction. While even advocates stopped short at recommendng the system for private homes, a few people built slow-burning houses. One house, a summerhouse in Yarmouth, Maine (1889–90), had plank walls as well as floors. Built by ship carpenters, it cost only 10 to 15 percent more than an ordinary frame house. H. W. Brown, former president of the Philadelphia Mutual Fire Insurance Company, also built a slow-burning house, in Philadelphia.[47]

The system became known to people outside New England. As early as 1879, Peter B. Wight, who undoubtedly learned about the system through the pages of AABN, brought it to the attention of New York insurance agents in a paper he read at a meeting. He distinguished between buildings that were first-class, or noncombustible, and second-class, or "practically fire-proof," and listed slow-burning construction as an example of the latter, describing it as a "heavy wood" system required "by the mutual insurance companies of Massachusetts and Rhode Island." A Philadelphia insurance leader and map publisher, Charles John Hexamer, brought word of the system to his region in an article he wrote for the *Journal of the Franklin Institute,* which repeated much of the information in Woodbury's textbook. Following a deadly fire in the Leland Hotel in Syracuse, New York, the magazine *The Architectural Era* recommended that "public buildings of all kinds . . . be built on fire-proof or slow-burning principles."[48]

Boston architects adept at the system brought it to projects outside New England. In Philadelphia, a group of investors that included H. W. Brown commissioned the Boston firm of Cabot & Chandler, with Amos J. Boyden, to design the Brown Building (c. 1882–83) on slow-burning principles. Theodore M. Clark, an architect and professor of architecture at MIT, used slow-burning construction for a loft building, the Lawrence Building (c. 1884), in New York City. Atkinson judged the firms of Peabody & Stearns and Shepley, Coolidge, and Rutan to be "thoroughly familiar" with the principles of slow-burning construction.[49]

By the early 1880s, even Western architects knew about the system. The Chicago firm Adler & Sullivan designed a huge factory for the Aurora

Watch Company in which "the eastern system of solid timber floors will be used."[50] Atkinson believed West's Hotel in Minneapolis to be an example of a slow-burning construction and that the system had been adopted by the state of Indiana for several hospitals for the mentally ill. Atkinson also considered the floors in Cupples Warehouse in St. Louis to be a model of the slow-burning type. Slow-burning construction became popular with Midwestern warehousemen, especially in St. Louis.[51]

The stock fire insurance industry took note of AFM's building rules as well as its methods of setting standards, inspecting properties, and educating owners. In 1886, the association of stock fire insurance companies in Boston (the Boston Board of Fire Underwriters) followed Atkinson's example and issued its own circular on safe construction, entitled "Slowly Combustible Buildings." Commenting on this pamphlet, the editors of AABN approved its intentions but criticized the specifics. The pamphlet's author, an inspector with the board, agreed that the circular could have been better written, but noted that it represented a first cut at formulating "the rules of 'Mill Construction,' now so universally insisted upon by all our great New England Mill Mutual Insurance Companies" for nonindustrial buildings.[52] Several warehouses in Boston, he wrote, were being completed according to his recommendations. Another example of the mutuals' growing influence was that a few stock insurance companies began to offer discounts when owners installed certain fire protection appliances, such as automatic fire alarms and automatic sprinklers.[53]

Around the mid-1880s, the reading public also had a chance to learn about the loss prevention approach and construction rules of the factory mutuals. An article in the *Forum*, with the very Atkinsonian title, "The Waste by Fire," praised the AFM approach of requiring owners to bring their properties up to standard before insuring them. *Scientific American* published Charles Woodbury's lecture on the evolution of mill architecture, in which he discussed the fire protection features of mills; a similar paper by Woodbury appeared in the first volume of *Cassier's Magazine*. The most extended discussion of slow-burning construction in a general-interest magazine was one written by Atkinson for the *Century Magazine* and published in 1889.[54] In the article, entitled simply "Slow-Burning Construction," Atkinson offered mill construction as a way to reduce the huge, and preventable, loss of property to fire each year. He introduced the concept of the "fire tax," which he asserted was "the heaviest tax imposed on the people of the United States."[55] It included the value of property destroyed in fires; the cost of fire insurance; and the cost of fire departments.

Other writers picked up the idea of fire as a tax and used it, for example, when arguing for stricter building codes.

This publicity convinced many in building circles that "slow-burning construction" would be a reliable alternative to fireproof construction. At an 1889 hearing on amendments to New York City's building law, a prominent developer urged the city to allow "slow-burning" buildings to exceed the proposed eighty-foot height limit for nonfireproof buildings. At the same hearing, the president of the Continental Fire Insurance Company, Francis C. Moore, testified that he would write policies on "slow-burning" structures at one-third off the rate for ordinary buildings. He even wrote a pamphlet about structural fire protection, *Economical Fire-Resisting Construction*, in which he quoted London's Captain Shaw on the fire-resisting capability of heavy timber construction.[56] As a leader in the movement to rationalize insurance rate setting, Moore's endorsement probably encouraged other underwriters to treat "slow-burning" buildings favorably. Around 1890, St. Louis's Board of Fire Underwriters reduced rates for owners of, and tenants in, buildings "constructed upon the slow combustion principle." Chicago's insurance board, like Boston's Board of Fire Underwriters, issued a circular defining slow-burning construction and offered lower rates on such buildings.[57]

Building codes, too, recognized slow-burning or mill construction as a fire-resistive system. Philadelphia's building law established four classes of building, the first being fireproof and the second, slow-burning construction. It allowed slow-burning buildings to have undivided floor areas three times larger than those permitted in ordinary buildings (fifteen thousand rather than five thousand square feet), and to be six stories or eighty-five feet tall, while ordinary buildings could not exceed sixty-five feet. Chicago's building law allowed slow-burning buildings to rise to a hundred feet while the limit for ordinary buildings was sixty feet. Similarly, Cleveland, New Haven, and San Francisco—all cities with height limitations— permitted buildings of mill construction to exceed the limit for ordinary buildings, because they were considered safer.[58] Surprisingly, the term *slow-burning construction* did not appear in Boston's code, but the principles did. The law of 1892 required floors in second-class buildings (first-class buildings being noncombustible) to be firestopped, meaning "covered with a solid, air-tight cohesive layer" of some fire-resistive material, except that plank floors could be substituted in certain cases. It forbade wood furring in both first- and second-class buildings, thereby eliminating hollow spaces; plaster had to be put directly on the inside face of walls or else on wire lath. The pitch of roofs on commercial or residential buildings over sixty feet tall could not exceed 20 degrees. Finally, it called for the ends of beams entering walls to be beveled to allow them to fall without disturbing the wall, as required in mill construction.[59] Some years later, a model building code for the nation recommended this treatment of beams, and building codes in many cities eventually required "fire cut" beams[60] (Fig. 4-2).

Thus, by the 1890s, the slow-burning system had found its way into a variety of building types around the nation, and had been recognized in insurance company rating schedules and building laws. Nevertheless, people frequently confused it with the semifireproof systems, and many designers used it inexpertly, usually neglecting to put in the sorts of fire suppression equipment invariably found in mills.

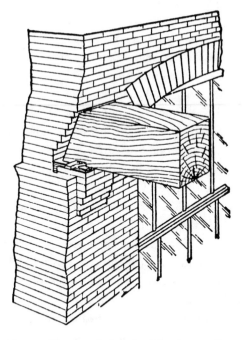

Fig. 4-2. Fire-cut beam. The beveled end of the beam allowed it to fall without disturbing the wall, should the interior of the building collapse in a fire. In this instance, the beam is simply held in place by a notch that fits over a small rib on the iron plate on which the end rests. Joseph K. Freitag, *Fire Prevention and Fire Protection*, 2nd rev. ed. (New York: John Wiley & Sons, 1921).

Failure of Slow-burning Buildings

A reason designers often applied the system ineffectively was that they had little guidance on how to how to adapt it to nonindustrial buildings. Indeed, the few publications about slow-burning construction, such a Woodbury's book on mills, discussed it only in relation to factories and warehouses. The section on mill construction in the first edition of a popular construction handbook, *The Architect's and Builder's Pocket-Book* (1885), merely repeated the material in Woodbury's book.[61] Nevertheless, architects designed hotels, office buildings, and schools in what they called the slow-burning system. The problems of adapting the system to buildings other than textile factories were not trivial. For example, in textile mills of the day, floors carried only about thirty pounds per square foot and at most sixty pounds per square foot, but warehouses had to support considerably more weight. Boston's and New York's 1885 building laws required warehouse floors to be designed to carry two hundred and fifty pounds per square foot. For such loads, a girder and beam floor (a longitudinal girder and closely spaced beams across the building's width)—the usual type in buildings on narrow and deep city lots—provided the necessary strength at a lower cost than a mill-type floor.[62] Atkinson's missionary work oversold the system, and many of the city buildings described by their owners as "slow-burning" would not have met his standards. As slow-burning commercial buildings proliferated, the chances that one would burn and bring disrepute to the system increased accordingly.

One notorious failure occurred November 28, 1889, when Boston suffered its worst fire since the conflagration of 1872. Called the Thanksgiving Day fire, it burned a swath through the same downtown area destroyed in 1872 and killed five firefighters, some of whom were crushed when the walls of burning buildings collapsed on them. The fire started in a building that Boston's Fire Marshal described as "partly 'mill-framing.'"[63] After the fire, AABN wrote that several of the buildings in the burned area had been designed to be fire-resistant, including architect Henry H. Richardson's celebrated, and reputedly slow-burning, Ames Building.[64] The daily press seemed to exult in the irony of the situation, running headlines such as "Fireproof Buildings Melt Like Wax in the Flames" and "Buildings Apparently Fire-proof as the Alps Completely Gutted." As one AABN reader fretted, "business men will quote to-days' newspaper accounts to prove the futility of erecting fire-proof structures."[65] AABN's editor responded that, in fact, none of the buildings in the burned district were fireproof and therefore *fireproof* construction had not failed. Rather, he said, this fire had been a test of the slow-burning system, and he concluded that it should be scrapped for city buildings. This view did not represent a change of policy for AABN, which had always regarded slow-burning construction as an expedient: useful until the time when lawmakers mandated true fire-resistive construction.[66] The Thanksgiving Day fire only reinforced the editor's reservations.

On reading this article, Atkinson dashed off a letter to AABN in which he asserted that if there were no fireproof buildings in the burned area, there were no slow-burning ones, either. He explained that, contrary to what the editor believed, the Ames Building was in no respect an "excellent example" of slow-burning construction: the only feature it shared with a slow-burning factory was plank floors. All its other characteristics—the unenclosed stairways; unprotected windows; lack of standpipes; lack of access to the building from the roof or the rear; and the light framing in its pitched roof—"were utterly inconsistent with the very elementary conceptions of slow-burning construction."[67] Atkinson then detailed the elements of a genuine slow-burning commercial building on a city lot, which was probably the first description of how to adapt slow-burning construction to a city building.

Nevertheless, many supporters of safer construction and stricter building regulation had begun to warn against the slow-burning system. However excellent it might be for cotton mills, it was not suitable for buildings in the congested parts of cities that faced the hazard of *conflagration*. Fires in cotton mills usually started inside, where they would be noticed by workers or a watchman, and dealt with quickly.[68] For closely packed city buildings, the main danger was fires on the outside spreading from building to building. It was because of this unpredictable hazard that AFM companies refused to insure city buildings. Even BMMFIC's own Charles Woodbury agreed that "slow-burning construction was not feasible for commercial buildings in cities" given their requirements for display and open floor plans.[69] Slow-burning construction might work for the sorts of nonindustrial buildings that were usually separated from others, like college or hospital buildings on spacious campuses. But in crowded cities, given the exposure hazard, any combustible material was vulnerable. Experts began to view slow-burning construction as a false economy.

By 1893, Peter Wight had come to regard it as such. He did not even discuss the slow-burning system when he reviewed American fireproof construction practice for an international architects' conference. Indeed, after describing two semifireproof systems, he warned "the words 'slow-burning' are not applicable to them, but refer only to buildings constructed with heavy floor-beams and thick plank floors, and not considered in this paper as fireproof in any sense." Wight explicitly differentiated slow-burning from semifireproof construction because people often called the latter type "slow-burning."[70] Wight felt indignant at the misapplication of the term to the tile-protected buildings he considered practically fireproof. By the same token, Atkinson fumed when the follies of "ignorant and incapable draftsmen" were passed off as "slow-burning."[71]

In 1892, the insurance trade magazine, *The Spectator,* published a series of articles, originally written for firefighters, that criticized fireproof construction and especially slow-burning construction, its "bastard sister." It quoted the fire chief of St. Louis saying that slow-burning features actually made buildings more dangerous: unlike ordinary buildings, mill buildings did not collapse in a fire but burned for "several hours, a standing menace to every building in the vicinity."[72] AABN referred to these articles when commenting on an 1893 fire in Boston that destroyed another supposedly slow-burning building. Boston was considering limiting the size of buildings with timber frames, and AABN's editors hoped such limits would encourage developers to build fireproof buildings.[73]

In response to this, Atkinson wrote a much-reprinted statement about slow-burning construction in which he attempted once and for all to correct the widespread misunderstanding that the system involved no more than plank floors. In "Mill-Construction: What It Is, What It Is Not," he explained that naturally timber would burn, "*if not suitably protected.*" By "protected," Atkinson did not mean a barrier, as in semifireproof construction, or anything structural, but rather the sorts of fire-extinguishing equipment—the automatic sprinklers, standpipes, pumps, and hydrants— and private fire brigades found in AFM mills.[74] Automatic sprinklers had become general in AFM mills by the 1890s, but, outside of a few theaters, were rare in nonindustrial buildings. Underwriters' boards offered lower rates for "slow-burning" buildings, but did not require them to have automatic sprinklers. Some architects appreciated the importance of fire suppression measures to slow-burning construction and urged underwriters to raise the rates on buildings that claimed to be slow-burning but lacked automatic sprinklers. The editors of *Fireproof Magazine,* an architecture magazine, warned that architects who put up commercial structures "in a manner similar to cotton mills of the New England States" but left out "all the excellent precautions for preventing fires which there prevail" were making a mistake for which they would "someday be held severely accountable by their own clients."[75]

Some of the severest critics of slow-burning construction came from the new field of insurance engineering, a group of self-made specialists that included architects, civil engineers, manufacturers, and fire insurance inspectors. By the 1890s, many of these men believed the day of halfway measures, like slow-burning or semifireproof construction, had passed and society must accept only fireproof construction for its city buildings. This was the founding philosophy of *Fireproof Magazine* (FM), which began monthly publication in 1902. FM printed technical articles about fireproof construction, descriptions of exemplary buildings, and analyses of noteworthy fires, along with much editorial comment. While the magazine itself could not have had a large circulation (it ceased publication in 1911), its contributors included some of the leading fire protection specialists, who were active in professional organizations, wrote books, and published articles in other periodicals. Thus, a much wider audience than the readers of FM knew its editorial position. None of the writers for this Chicago-based magazine came from the New England mill world.

The magazine advocated steel and tile fireproof construction and considered "slow-burning" to be nothing of the kind. William Clendenin, its editor in 1903, regularly published derogatory comments about mill construction:

> The denuding of our forests brings at least one consolation. There will be less mill construction.
> The merit of mill construction is an imputed merit. Good authority won't put out a fire.
> The wooden Indian of mill construction, with painted and expectant face, hand extended—palm up—is still at a *premium*—with underwriters.
> East Pepperell, Mass., has on view some very interesting exhibits of what mill construction is *not*. What it *is*, is ruins, in East Pepperell.[76]

Fire protection insiders knew quite well these barbs were aimed at Edward Atkinson: "Good authority won't put out a fire" meant, trust the ideas of the AFM underwriter at your peril! Nor did FM shrink from criticizing Atkinson by name. F. W. Fitzpatrick, an architect and fire protection expert, commented unfavorably on one of AFM's reports on slow-burning construction and charged that self-interest motivated Atkinson to cling to outdated ideas:

If he feels that his best days are not yet over, that he can still learn and be useful, why does he not throw what influence he retains in the direction of really good construction? . . . On the one side of the scale lay these considerations: public safety, real economy, safe investment, protection to life and limb, and permanency of construction; on the other side place these considerations: 85 percent dividends in the New England Mutual Association and the usual commission upon the sale of automatic sprinklers. Which side will Mr. Atkinson choose?[77]

Some fire protection authorities—in addition to the New England mill designers—rejected the idea that fireproof construction was the only true way. One of these, Peter McKeon, wrote that it was idle to talk of fireproof buildings when the real danger to cities lay in the standing stock of fire-traps. The challenge was to make the ordinary buildings safer, which could only be done through careful watching and retrofit, including installing sprinklers and alarms; isolating stairways and elevators; protecting windows; and organizing fire drills. In short, taking steps to improve the existing building stock would do more to reduce the risk of conflagration than the occasional fireproof building. Not only was this course more practical, he wrote, but it made more sense on cost-benefit grounds: money wasted for a good cause like public safety was still money wasted.[78]

Nor could Atkinson be fairly characterized as a stubborn defender of old ways and indifferent to new methods. On the contrary, he spent his last years trying to raise funds to establish an insurance engineering course at MIT or, failing this, to start a fire protection research laboratory there. Atkinson did raise enough money to set up the Insurance Engineering Experiment Station (IEES) at MIT in 1902. The laboratory studied the fire resistance of new building materials, including reinforced concrete, which had been little used in the United States, and a roof made of corrugated steel and cement.[79] The lab's positive findings about concrete turned Atkinson into an early proponent of reinforced concrete for factories. The editor of *The Cement Age*, one of the first American magazines devoted to concrete construction, prominently reported his views.[80] Right up to his death in 1905, Atkinson worked for causes that he believed would improve public well-being. In an obituary in FM, Peter Wight—for a time FM's editor—publicly recognized Atkinson's important contribution: he was, Wight wrote, fire prevention's "most conspicuous prophet."[81]

Meanwhile, buildings advertised as "slow-burning" burned. FM reported on a fire in West's Hotel in Minneapolis in 1906, a building that Atkinson twenty years before had cited as a good example of slow-burning construction. Ten people died in the blaze. Atkinson could not have known the construction details of this circa 1880 hotel, which was actually semi-fireproof and lacked sprinklers and a water supply for its standpipes. Even so, the structure contained the fire, preventing it from spreading. The high death toll was due to nonstructural defects: the absence of a general fire alarm system and badly designed exits.

Mill designers continued to use the slow-burning system in the early twentieth century, but new conditions made wooden mills less desirable than alternatives. For one thing, many manufacturers found their plants encircled by cities and their own factory yards crowded as they put up new buildings and enlarged old ones. This increasing density made the sites vulnerable to sweeping fires, something slow-burning construction could not protect against. Reinforced concrete, which began to be used on a commercial scale in the first decade of the new century, offered several advantages over wood, in addition to noncombustibility. A concrete building could be stiffer than a wood-framed one, which reduced wear on machines. Concrete also would not rot or decay, and its comparative impermeability was an advantage for wet operations—for example, in paper mills, dye works, and bleacheries—and food products manufacturing. Reinforced concrete permitted much longer clear spans than wood construction: by 1907, concrete beams had been built with unsupported spans as long as fifty feet. Last but not least, a factory built of concrete cost only a little more than a slow-burning one. The size of the difference varied according to the region—greater in the South, less in the New York City area—and the details of a building. But in New England and the New York region, the cost premium for a reinforced concrete factory was no more than 10 percent, and the factory would be fully fireproof. The prices of materials of the slow-burning mill were rising, too: large-dimension, well-seasoned timber beams from certain species of trees (i.e., the preferred type, yellow pine) were becoming increasingly expensive. All these factors combined to make the multistory, wooden, slow-burning mill obsolete. The AFM companies discontinued issuing their slow-burning construction standards—long published as Report No. 5—in 1925.[82]

A Legacy

Although slow-burning construction became obsolete, the system left an important legacy. Designers of fireproof buildings typically relied on the noncombustibility of the materials to safeguard the structure. But in slow-burning mills, because they were not fireproof, designers introduced thorough compartmentation and internal means for fighting fires. The fire safety of American buildings improved when these mill measures were combined with noncombustible construction.

Perhaps the most important fire protection item the mills bequeathed was automatic sprinklers. Automatic sprinklers could spray water on fires in hard to reach places, could continue flooding an area after people had to flee, and could start working even when no one was around. Moreover, they could be added to existing buildings. The early perforated pipe sprinklers had two principal drawbacks: they required people to turn them on and they flooded the area all along their length. An early solution to the second problem was to install separate rising pipes for each floor of a building, each with its own valve, and labeled with the name of section of the building it served. Still, the valves had to be turned on by hand.[83] English inventors patented several Rube Goldberg-like devices with heat-sensitive triggers that, when set off, opened valves and released water into pipes.

The first practical automatic device contained an important innovation, a "head" that screwed into a threaded opening in the sprinkler pipe and sprayed water when the solder that sealed it melted.[84] This allowed the pipes to be kept filled with water under pressure. Not only was the system "automatic"—in that the fire itself activated the sprinkler—but only the heads near the fire would open, so water would be concentrated where it was most needed. A final bonus: the flow of water into the pipes activated a fire alarm. Stewart Harrison, an English firefighter, developed the system in 1864–65 and chose not to patent it. A contemporary published information about the system and recommended it for ships as well as "churches, libraries, halls, museums, & c." Yet Stewart's countrymen did not take up the idea. In England, before 1890 at least, sprinklers were rare, even in mills. British fire insurance companies, concerned about leakage and water damage, discouraged their use and charged higher rates for sprinkler-equipped buildings.[85]

In 1874, Henry Parmelee of New Haven patented the first practical American automatic sprinkler. Parmelee invented two styles of sprinkler heads; the more successful of the two had a cap held in place with solder and came to be called the "sealed" or "water-joint" type head. In the layout of the pipes—one running the length of a floor, with branches going into each bay—Parmelee's plan resembled Harrison's. In 1875, a Fall River, Massachusetts mill owner became the first AFM member to adopt these sprinklers, not long after a fire burned one of his mills, killing twenty-three workers. He was more receptive than other mill owners, many of whom distrusted automatic appliances.[86]

In 1878, Parmelee arranged for Providence Steam and Gas Pipe Company (PSGPC), manufacturers of perforated pipe sprinklers, to make and install sprinkler systems using his heads. He also patented a new sprinkler head that had a little turbine inside, which sprayed water from slits and doused a wider area. In the field, this new sprinkler head responded slowly because water flowing into the head at first cooled the solder and resealed the head briefly. A solution to this problem was the "sensitive type" head, patented in 1879, which placed the soldered joint away from the water. In 1881, PSGPC's president, Frederick Grinnell, introduced another sort of "sensitive type" head with a metal disc valve, which was held in place by a lever soldered at the bottom of the head. This head introduced another important new feature: a bottle cap-shaped piece of metal, called a "deflector," that forced water out over a wider area (Fig. 4-3). In the first ver-

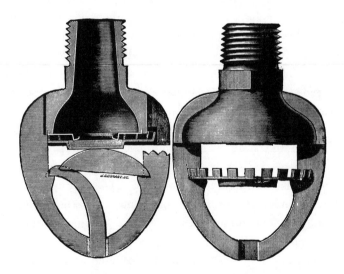

Fig. 4-3. Early Grinnell Metal Disc Sprinkler head, circa 1881. Drawing on the left shows a section through a closed head. On the right, an outside view of head when opened. A rocker and lever pushed against a soft metal disc, which plugged the opening. When the solder holding the lever melted, both it and the rocker fell away, and water forced out the disc. The bottle cap-shaped piece created a spray. Grinnell issued several improved versions of this type in the 1880s. Charles J. H. Woodbury, *The Fire Protection of Mills* (New York: John Wiley & Sons, 1882).

sion of the disc style of head, the soft metal disc, under pressure from the lever, tended to deform and stick in the opening. Grinnell redesigned the head to correct this problem and continued to improve it through the 1880s.[87]

By 1880, sprinklers, manual or automatic, were universal in the picker and drying rooms of AFM member's plants, but less than a third of their main mills had them. Edward Atkinson became convinced that mills, too, needed sprinkler protection, and therefore in 1881 he set up a new mutual company, Spinners Mutual Fire Insurance Company, to offer reduced rates on mills fully protected with sprinklers. Soon, the AFM companies agreed to require sprinkler protection in mills; however, they did not require *automatic* sprinklers, which were still more or less experimental devices in the early 1880s. In order to help members choose among the various automatic sprinklers available, AFM, with assistance from the Boston Underwriters' Union, conducted tests of many models of sprinkler heads from ten different companies, as well as old-style heads removed from mills, to determine their reliability. The tests covered such features as durability, tendency to leak, ability to distribute water, and cost, as well as how well they extinguished fires. The tests uncovered defects in various heads, which their manufacturers then corrected.[88]

Sprinklers continued to evolve. In 1890, Grinnell introduced a head in which he replaced the metal stopper with a glass button. It incorporated features invented by BMMFIC engineer John Freeman, which he sold to Grinnell. This sprinkler head became the standard one for several decades and the ancestor of modern sprinkler heads. Another advance, at the turn of the century, was the introduction of dry pipe sprinklers, in which water was held back by compressed air. These could be used in places where water-filled pipes might freeze in the winter. Although the new heads improved on the old, the older models continued to perform effectively. A comparison of the value of property damaged in fires in rooms equipped with automatic sprinklers and those without, between 1877 and 1891, showed that the average loss per fire in the former was only 7 percent of the latter.[89]

Yet even with AFM's strong endorsement and financial inducements— lower rates on sprinkler-equipped mills—automatic sprinklers spread slowly through the mill world in the 1880s and 1890s. In the early days, paradoxically, AFM's already low insurance rates depressed sales among their customers. Besides doubting that automatic sprinklers would work when needed, some mill owners feared they would open unexpectedly and cause water damage. Therefore, some owners of mills with automatic sprinklers took out insurance against water damage, although they could not get this coverage from the AFM companies. Atkinson lobbied to change Massachusetts's insurance law to allow BMMFIC to offer leakage insurance. BMMFIC wrote separate leakage policies from 1895 until 1902, when it could include water damage in its regular policy. In 1910, AFM companies mandated automatic sprinklers for all the properties they insured.[90]

Sprinklers made almost no headway outside factories, although fire safety experts recommended them for hazardous buildings such as theaters, department stores, and hotels. In 1877, the year after the terrible Brooklyn Theatre fire, which killed nearly three hundred people, a group of architects drafted a theater law for New York City that required "a thorough system of perforated pipes or sprinklers to be provided both on the stage and in the auditorium."[91] Two years later, in an article about theater design, a Boston architect suggested furnishing the area over the stage with "a net-work of perforated pipes known as a 'factory sprinkler,'" and placing a conspicuous turn-on valve near the prompter's box.[92] Sometimes mill owners brought sprinklers to commercial buildings: a theater in Woonsocket, Rhode Island, installed automatic sprinklers at the suggestion of one its investors, who managed a cotton mill.[93]

Why did it take so long for sprinklers to spread to buildings besides factories? For some years, sprinkler companies did not even try to sell their product to nonindustrial customers. For example, no sprinkler manufacturer advertised in AABN in the 1880s. Sprinklers were expensive to install, probably more so in a nicely finished nonindustrial building than in an unadorned mill. Also, in the early days, at least, the stock fire insurance companies did not reduce premiums on properties that had them.[94]

Building laws created new customers for sprinklers. Theaters were the first kind of building in which sprinklers were compulsory: New York City's new building law of 1885 required automatic sprinklers over and around the proscenium opening and at the ceiling over the stage (but not in the auditorium). Boston's 1885 building law, like New York's, for the first time contained detailed rules for theater construction, calling for a perforated pipe over the proscenium opening and any additional perforated pipes or automatic sprinklers "as the inspector shall direct." In an 1886 summary of Charles Woodbury's report on sprinkler heads, AABN stated that "all the new first-class theatres in this country, we believe without exception, are equipped with a full sprinkler service over the stage."

By 1892, automatic sprinklers in theaters had helped douse a few fires. In 1905, about 150 theaters in the United States had sprinklers.[95]

Sprinkler sales got another boost in the 1890s when the stock fire insurance industry became interested in them. In Cincinnati, insurance companies offered discounted rates on buildings with roof water tanks and automatic sprinklers. In 1896, an association of stock companies, the New England Insurance Exchange, approved the sprinklers of ten manufacturers. New York City's Board of Fire Underwriters became sprinkler converts in the early twentieth century and urged their adoption. One prominent fire protection authority, Joseph K. Freitag, even recommended them for fireproof buildings, to protect the contents of the building in a fire. Sprinklers could tackle fires on the upper floors of high buildings, which fire departments had difficulty reaching. Several destructive fires in tall buildings proved the need for supplementary protection.[96]

Because they originated and developed in New England, the region had more sprinkler-equipped buildings than any other by the early twentieth century. By one estimate, a third of all property insured by fire insurance companies in New England (measured in value) was protected with sprinklers. In Greater New York, by comparison, in 1905, only 605 sprinkler installations had been made.[97] Eventually sprinklers came to be required by law for a variety of buildings. But the early development and spread of this important fire suppression device was largely due to the encouragement and technical assistance given by the AFM companies.

The AFM's loss prevention approach—which included setting and enforcing standards for construction, equipment, and maintenance of members' properties—had saved its members a great deal on their insurance costs, not to mention danger, trouble, and loss from fires avoided. As word of AFM's success spread, manufacturers in other regions organized companies modeled on the New England mill mutuals. In Philadelphia, the Philadelphia Manufacturers' Mutual Fire Insurance Company formed in 1880. Later the Western Manufacturers Mutual in Chicago, Central Manufacturers Mutual in Ohio, and Manufacturers and Merchants in Rockford, Illinois, were established. Atkinson urged manufacturers in industries that experienced large fire losses—and thus paid high insurance rates—to organize mutuals among themselves to improve their safety. One group that consulted Atkinson on this was New England rubber manufacturers. The insurance trade paper, *The Spectator,* reported that when they learned what it would cost to install necessary safeguards, their enthusiasm "disappeared, and the stock companies are likely to collect premiums of them for some time longer."[98] But this prediction proved wrong: they formed a mutual, the Rubber Manufacturers' Mutual Insurance Company of Boston, that very year, 1884. Atkinson believed the stock companies exaggerated the risk of paper mills and therefore overcharged to insure them. Soon, the Paper Mill Mutual Insurance Company of Boston started, with Atkinson as president.[99]

This movement alarmed the stock fire insurance companies, who saw the mutuals skimming off the best industrial customers. To compete, in 1890 a group of stock companies organized their own specialized insurance pool, the Factory Insurance Association (FIA), and offered rates comparable to those of the AFM companies for similar classes of properties. Atkinson accused FIA of setting prices below what it could afford to charge, in order to steal away AFM's members. Nevertheless, the FIA intended to have low losses because it planned to follow AFM's methods. For its first manager, it hired Francis W. Whiting, a son of AFM's chief inspector, William Whiting.[100]

Owners of nonindustrial properties often applied to AFM companies for insurance, but were turned away. Atkinson urged owners of like properties, such as churches, theaters, and colleges, just like manufacturers, to organize their own mutual insurance companies. In 1883, he proposed a mutual system for city commercial buildings, which he believed would encourage them to install protective devices such as standpipes and roof hydrants. In fact, specialized mutual companies did start up in the last two decades of the century. Wisconsin's mutual insurance law authorized companies to insure a variety of kinds of buildings; church mutuals were especially popular there.[101]

The movement to form mutuals had more to do with public distrust of fire insurance companies than a dedication to reducing fire waste. Americans had a variety of complaints about the insurance industry, but mainly they believed that insurance rates were too high or inequitable. Responding to public dissatisfaction, state legislatures enacted laws aimed at controlling rate setting and other insurance company practices. This riled the insurance industry and it was to derail such legislative meddling that the companies attempted to prove their current rates were fair and reasonable, or else to reduce them. But to reduce rates while continuing to operate profitably required reducing costs, especially losses. In the 1890s, a reform group in the stock fire insurance industry urged their colleagues to do just this: to take steps to bring down fire losses and share the resulting savings with customers in the form of lower rates.

Some stock insurance companies began to take steps to reduce losses. In a number of cities, the local underwriters' associations, which represented the insurance companies doing business there, hired engineers to inspect special hazards such as electrical systems, sprinklers, and boilers. Boston's and Philadelphia's underwriters' groups established an inspection service and, according to Edward Atkinson, "have engaged two of our best men to set the work going." The Underwriters' Bureau of New England published guidelines for construction, which incidentally included rules for "standard mill buildings."[102] In Chicago, William H. Merrill, a young electrician from Boston, proposed to the local underwriters' group that it support the establishment of a research laboratory to analyze the causes of electrical fires. Merrill had studied electrical engineering at MIT, and probably learned about AFM's research laboratory during his college years and later while working for the Boston Board of Fire Underwriters. The Chicago underwriters' board financed his project, and some years later the National Board of Fire Underwriters (NBFU), a trade association for the stock fire insurance industry, became the main backer of the testing lab. Merrill expanded the work of the lab to include testing of a range of devices and appliances that had a bearing on fire safety and in 1901, incorporated it in Illinois as The Underwriters' Laboratories. Meanwhile, representatives of stock fire insurance companies in the East joined to establish uniform rules for sprinkler installation. After accomplishing this, the group turned its attention to other fire safety matters and then organized the National Fire Protection Association, "to promote the science and improve the methods of fire protection and prevention."[103]

Most stock fire insurance underwriters considered individual building fires good for business; but conflagration could cause their ruin. Therefore, they lobbied cities to invest in fire fighting (better water systems and fire departments) or enact certain building regulations, and threatened to charge higher rates if the cities did not. At the turn of the century, to check the losses they might suffer in case of conflagration, the large companies set limits on the total value of property they would cover in city centers.[104] Owners therefore might have difficulty obtaining coverage in cities the underwriters judged unsafe. But in the cities they considered safe, insurance companies were willing to do a larger amount of business and the increased competition drove down insurance premiums. The prospect of lower premiums convinced many in the industry that the reformers' fire prevention strategy would hurt business by squeezing income. The reformers countered that lower rates were necessary to soothe an unhappy public and ward off meddling legislators. In order for companies to reduce rates generally and stay profitable, cities had to be safer.

In the spring of 1904, following the devastating conflagration in Baltimore, the reform forces persuaded the National Board of Fire Underwriters to sponsor comprehensive studies of the fire safety of the nation's large cities, in order to evaluate the conflagration danger in each one. Members could use the resulting information to prepare more realistic rate schedules. A Committee of Twenty was appointed to direct the survey work, with Henry Evans, president of Continental Insurance Company, as chairman. By the end of its first year, the committee assembled a technical staff that included forty-two hydraulic, mechanical, electrical, and civil engineers and support staff. They went out in teams to cities around the country, and their reports did more than inform underwriters about fire safety conditions: they also served as guides for city officials. When cities took the advice in the studies and upgraded their water systems, fire departments, and building rules accordingly, they reduced the likelihood of widespread fire, which benefited local residents as well as the companies that insured their property.

Yet many underwriters still believed that they had no business working to reduce fire danger and objected to the very high cost—$104,500 for the first two years—of the city survey project.[105] The competing philosophies of the traditionalists and the reformers collided at the 1905 annual meeting of the NBFU and reveal how influential the mill mutuals had become. Having gathered a group of highly qualified technical men, and with the survey process in full swing, Henry Evans begged the National Board to continue funding the project. The work, he argued, showed the public that stock underwriters cared about the harm fires caused, not only about collecting premiums. Evans predicted that if the stock companies made no effort to bring down fire loss in cities, city merchants—some of whom invested in textile mills and therefore were acquainted with the mutuals—would turn to the mill mutuals for help.[106]

The president of a large New York insurance company who opposed continuing the project responded that the city reports were too detailed and cost too much to prepare. Why not assign this survey work to the NBFU's Fire Department Committee, which could do it for less money, he asked? At this, two representatives of insurance companies in Providence, Rhode Island—a center of mill mutual activity—rose to warn their colleagues what they could expect if the mill mutuals decided to pursue commercial customers. Underwriters from outside New England who did not have to compete regularly with the mutuals could not imagine the derision stock company officials faced from businessmen who insured with the mutuals. From the businessman's point of view, the mutuals had a system and engineering expertise, while the stock companies did not. The reason the stock companies would lose in competition with the mutuals, they warned, was precisely because customers believed the mutuals served the interests of the customers, by helping them reduce losses and prevent fires. For these two underwriters, the Committee of Twenty's report for the city of Providence was the first thing they could show to local businessmen to prove that the stock companies had technical expertise and wanted to help their customers. The business leaders in Providence welcomed the report, and the city administration voted funds to upgrade the fire department and water system. This was exactly the sort of effect the survey project was intended to have, and the two men urged the board not to be penny-wise and pound-foolish.[107] The reform forces prevailed. The stock insurance industry began to make loss prevention and safety part of its goals.

The 1904 Baltimore conflagration helped turn the stock fire insurance industry toward loss prevention, but it also brought new discredit to the name of slow-burning construction. The fire began in the warehouse of Hurst & Company, which some contemporary accounts said was of mill construction. From a description of the building, this seems doubtful, but such reports tarnished its reputation nonetheless.[108] One observer, a writer for *Engineering News*, saw a vindication of the slow-burning system in the aftermath of the fire. He noted that while the conflagration had reduced buildings to heaps, the heavy timber telegraph and telephone poles along the streets of the burned area survived. This, he wrote, showed "the value of the so-called 'slow-burning' mill construction."[109]

TRIUMPH

The Fireproof
Skyscraper

Without steel buildings the present science of fireproofing would
never have been called into existence, while without the development
of fire-resisting principles steel construction, as applied to buildings,
must have been discontinued long ago.

JOSEPH K. FREITAG, civil engineer and fire protection expert, 1906

Compared to its fiery start, the 1870s ended quietly, and the following decade opened peacefully, from a conflagration standpoint. But the nation's
fortune turned as the decade progressed and climaxed in the terrible year
for fires, 1889, when Boston's downtown burned in the Thanksgiving Day
fire and sweeping fires struck New York City; Seattle and Spokane Falls in
Washington; and Lynn, Massachusetts. By one estimate, just eleven blazes
in this year destroyed $30 million in property.[1] It seemed fire waste was
bound to increase, as the downtowns of large cities expanded and filled
with bigger, more costly buildings.

One of the spurs to downtown development in the 1880s was a tremendous demand for office space. Office buildings went up in cities around the
country, and many of them were fireproof. Two interrelated factors accounted for this growth in fireproof construction: the unprecedented
heights of some of the new office buildings and, out of concern for the fire
safety of these tall buildings, laws that for the first time required tall
buildings to be fireproof.

Fire Safety of Tall Buildings

Improvements in the safety and reliability of passenger elevators in the
1880s allowed developers to breach the practical six-story limit for walk-
up buildings. Elevators rearranged the economics of office buildings: in
buildings without elevators, rents declined with height, while in an elevator building, the upper stories fetched higher rents than the intermedi-

ate ones did (the first two floors still commanded the highest rents). The high buildings in this decade rarely exceeded ten to twelve stories, apart from towers; nevertheless, these were unheard-of heights. It was to describe the awe-inspiring impact of the ten-story buildings that someone coined the term *skyscraper.*[2]

The upper floors of the high office buildings, remote from the noise and dirt of the street, appealed to many tenants, but some worried about the security of these buildings in case of fire. Firefighters had difficulty putting out fires high above the ground. Tenants feared losing their property—business records, special equipment, drawings, and so forth, which they could not easily replace—to fire. Given the demand for secure buildings, some investors tried to make their buildings at least appear to be safe. The owner of Chicago's first skyscraper, the ten-story Montauk Block (1881–82), did not plan to make it fireproof, but still wanted the public areas finished in a way that would reassure tenants. Thus, he ordered the lobby walls to be made of "face brick with red or black mortar (if as cheap as plaster), which would convey the idea of 'fireproof' to the whole structure—a valuable idea in a building of eight stories." But before construction began, he decided to make the building fireproof. It was more usual for owners to pretend their buildings were fireproof, since the average commercial tenant could not know the difference.[3] Many owners passed off their semifireproof buildings as "fireproof."

Moreover, before the mid-1880s, owners had no legal obligation to build fireproof. In only one city—New York—did the building law even recognize fireproof construction as such. New York City's 1860 law contained specifications for fireproof construction, but required only one kind of building to be fireproof: large tenement houses that lacked either fireproof stairways or fire escapes.[4] The likelihood of anyone actually building a fireproof tenement was nil, and this rule disappeared in an 1862 revision of the law. In 1867, a new rule again called for tenements to be fireproof in another rare case: when two houses were to be erected on a single lot, front and back.[5] This law also required stairways inside large tenements to be fireproof. In 1871, lawmakers dropped the fireproof tenement rule but did order tenements to have fireproof floors over ground-level shops or storage rooms.[6] Thus, in New York in the 1860s, tenements were the only kind of building legally required to be wholly or partly fireproof; lawmakers did not institute any safeguards for tall buildings as such.

In the 1880s, when the first tall buildings appeared in cities around the United States, residents often regarded them with pride, as manifestations of their city's vitality. But as their numbers increased, especially in Chicago and New York, many people voiced apprehension and even outright disapproval of this proliferation.[7] It was in these cities that the movement started to restrict building heights. Public opinion in Chicago tipped when buildings reached twelve and thirteen stories, at which point the possibility of a downtown packed with tall buildings became more real. People worried that tall buildings would block the light and air of their neighbors; concentrate people and thereby cause traffic and pedestrian congestion; and drive up land prices in certain districts while other areas languished.

Fire safety, too, was a major concern. Given ordinary water pressure and fire-fighting equipment, the highest that water could be sprayed from ground level was about eighty feet. Firefighters on the ground could fight fires only up to sixty feet with any effectiveness. A few commercial buildings had standpipes with outlets and hoses for fire fighting, but in the vast majority that did not, firefighters had to drag their heavy hoses and gear up the stairs. The open stairways commonly found in tall buildings were not designed to help firefighters get to a fire or the occupants to escape. Moreover, occupants could not be rescued from the upper floors of a tall building once the stairways filled with smoke and flames, since no fire ladders of the day reached great heights. Elevators, usually in open shafts, acted like chimneys and spread fire from floor to floor; they were of little use for access or egress in a fire.[8]

Unlike in the United States, many European cities imposed limits on the height and floor area of buildings to reduce fire hazard. On the Continent, height restrictions served esthetic as well as safety objectives—a means to create a uniform street wall. Paris limited heights according to the width of the street. London at first was more lenient. The Metropolitan Building Act of 1844—that city's first comprehensive code—controlled the heights of buildings on newly laid out streets, but not those of new buildings on existing streets. It capped the overall size of warehouses at no more than 200,000 cubic feet between party walls, a rule that affected height. But in 1855, a new Building Act set an upper limit for all buildings at 100 feet; it also increased the allowable volume for warehouses and factories slightly, to 216,000 cubic feet.[9] The law did not require any building to be completely fireproof; however, it did require fireproof construction in the exit ways—lobbies and corridors—of public buildings and apartment houses.

Americans who opposed the spread of high buildings wanted their cities to adopt European-style height restrictions. Initially, however, American cities took a different route to controlling the hazard of tall buildings: rather than limiting size, they required that buildings over a certain height be fireproof. Chicago was the first city to enact such a rule, in response to pressure from the city's fire underwriters.[10] Chicago's 1884 ordinance called for buildings over ninety feet to be "built throughout of incombustible material."[11] The following year, Boston and New York City adopted new building laws and both required tall buildings to be fireproof. In Boston, buildings over eighty feet, excluding churches and grain elevators, were to be "constructed throughout of incombustible materials, excepting interior finish." New York City set a lower threshold: new buildings exceeding seventy feet "shall be built fire-proof." But two years later, lawmakers relaxed the rule a bit, raising the cap to eighty feet.[12]

This growth of tall buildings had important consequences for the safety of downtowns, the development of the fireproofing industry, and architectural practice. If tall buildings were fireproof, then the more a city added, the safer (more fire-resistive) it became. It was through this process of redevelopment—with fireproof buildings replacing ordinary combustible ones—that cities finally were able to conquer urban conflagration. At the same time, the tall buildings expanded the demand for fireproof construction materials, stimulating invention and transforming what had been specialty products into commodities. Finally, designing the fireproof buildings trained architects in the new technology. Before the 1890s, a limited number of architects received most of the commissions for fireproof buildings. But with building codes requiring that tall buildings as well as theaters, roofs, and, in New York City, even tenement house floors be fireproof, many more designers had occasion to learn about fireproof construction.

Architects turned to iron and steel manufacturers' handbooks for guidance on how to detail fireproof buildings. The handbooks of the rolling mills evolved from catalogues of products to condensed textbooks, with illustrations of designs for fireproof floors and roofs (Fig. 5-1). The engineer C. W. Trowbridge advised architects, when designing structural metal, to consult "the valuable hand book of the Dearborn Foundry Company," because it "gives perhaps the most complete data on the subject; this book should be in the hands of all draftsmen."[13] In 1888, Charles Woodbury, the AFM fire protection engineer, attributed the increased use of rolled iron in large part "to the excellent and reliable engineering information contained in the manuals and catalogues issued by the rolling mills."[14] Paul

Starrett, a prominent builder, recalled that in the early days of skeleton frame construction, draftsmen relied on the mill handbooks to figure the sizes of steel members, which he believed resulted in much "amateurish engineering."[15] Similarly, hollow tile manufacturers published their own literature that covered the design of fireproof floors. In 1885, the Pioneer hollow tile company put out a handbook entitled "Fireproofing," with information on how to use their products.[16]

The Development of the Skeleton Frame

After the mid-1880s, when several cities passed laws requiring that tall buildings be fireproof, the course of the development of the tall building and the fireproof building became closely linked. In traditional masonry structures, walls carry the loads of the floors and roof, and the heavier the load, the thicker the walls must be. Except in small, lightly loaded buildings, the walls taper, because the lower floors support more weight than the upper ones, and therefore wall thickness at the base must increase with height. City building codes might require walls to be thicker than they had to be for structural stability alone, as a fire safety measure. Thicker walls were more likely to contain a fire and stay upright should a fire gut the interior. New York City's 1885 building code called for the walls of ten- to eleven-story buildings to be twenty-eight inches thick at ground level, and twenty-four inches thick up to about the sixth or seventh story.[17] But thick walls had several disadvantages from the owner's standpoint. The thickest parts would be in the lower stories, taking up valuable space. Thick walls also created deep recesses for windows, which blocked natural light. They added to the cost of construction and increased a building's weight, which was a particular problem for sites with weak soil and distant bedrock, like those in Chicago's business center.

At the turn of the 1890s, a new method of constructing high buildings evolved that overcame these drawbacks. Called "skeleton" or "skeleton frame" construction, the system differed from wall-bearing construction in that a metal—iron or steel—framework, rather than the walls, carried the weight of the floors and roof. In fact, the frame also carried the walls. Relieved of their traditional structural role, the walls of skeleton buildings could be a mere curtain: those at the lowest stories had to be no thicker than the one at the top floor. The system revolutionized methods of building construction.

The skeleton system was a logical, although not inevitable, outgrowth of improvements in the materials and methods of constructing tall, heavy

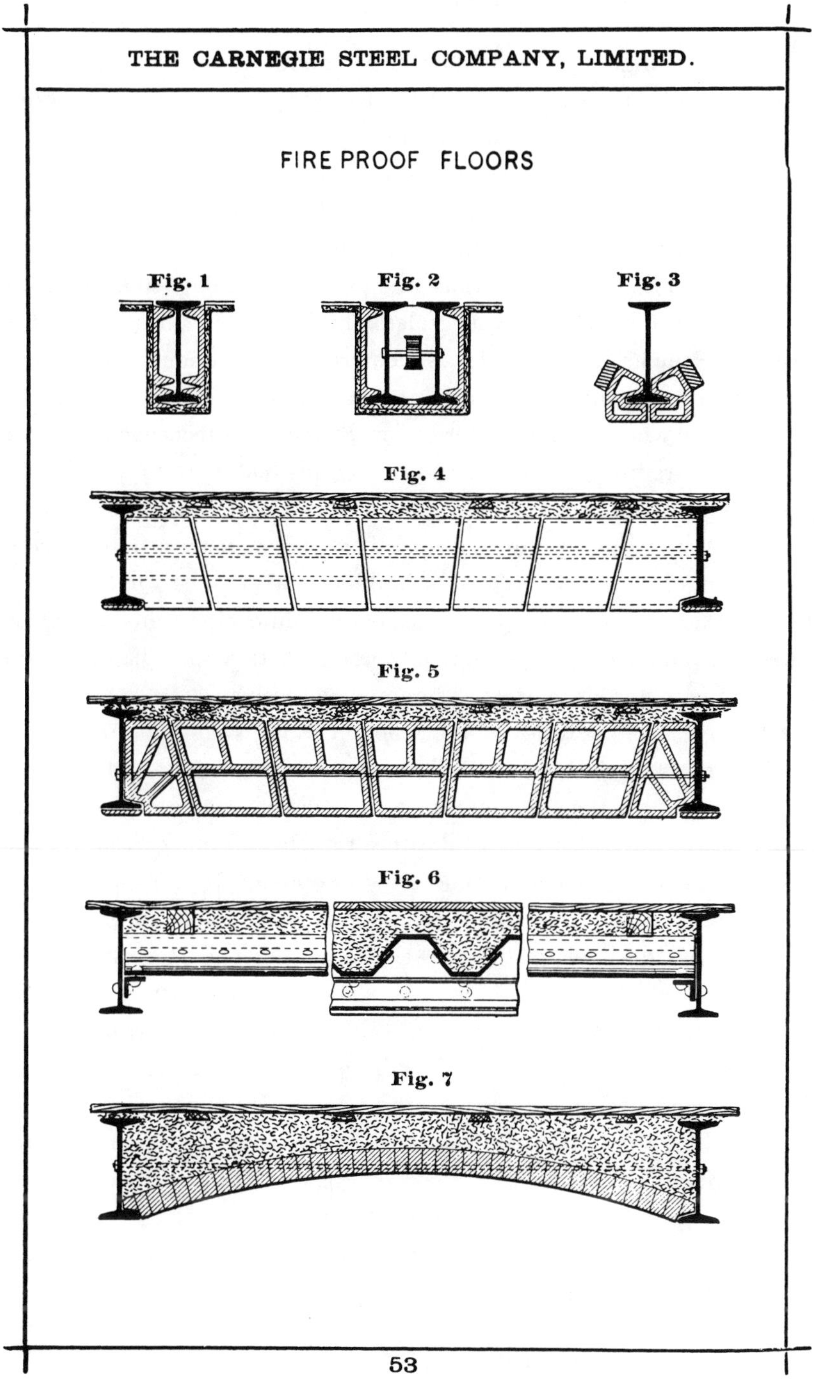

FIRE PROOF FLOORS

Fig. 1   **Fig. 2**   **Fig. 3**

Fig. 4

Fig. 5

Fig. 6

Fig. 7

Fig. 5-1. Fireproof floor details from an 1896 Carnegie Steel Company *Pocket Companion*, a handbook issued by the mill. Carnegie Steel Company, *Pocket Companion*, 1896.

fireproof buildings. The technological preconditions for creating a skeleton frame building existed by the end of the 1870s. The physical ingredients included structural iron for columns, beams, and girders; materials for making noncombustible roofs; and hollow structural units (tile or concrete blocks) for making floors and interior partitions. But the mere existence of these materials did not produce the skeleton frame. Western European designers also had these materials, and at an earlier date than American designers, yet they did not create skeleton buildings. One important difference: European architects used these materials less frequently. Nowhere in Great Britain was any sort of building legally required to be fireproof. Except for those who designed factories, few British architects had experience working on fireproof buildings. America simply had more fireproof buildings. The enthusiasm of real estate developers for tall buildings at the end of the century, which had to be fireproof, created opportunities for a growing number of American architects, builders, and engineers to learn about and use the new technology. The skeleton frame represents one of the many innovations they developed in their efforts to reduce the cost of, and improve the materials in, the tall, fireproof building.

Although some histories of the skeleton frame system claim cast iron facades, or structures such as iron greenhouses and the Crystal Palace, to be the ancestors of the skeleton frame, this view is based more on external appearances than structural similarities. Cast iron fronts carried no load; greenhouses and the Crystal Palaces of London and New York essentially were large roofs, not multifloor buildings. A more likely ancestor of the wall columns in a skeleton frame was the masonry pilaster wall. The pilaster wall used the same amount of masonry as an ordinary wall, but concentrated it into heavy, bearing parts, called piers, which carried the floor beams. This arrangement allowed for larger windows, since the panel between the piers had no load-bearing duty. Some of the early iron and brick fireproof buildings, from the 1850s, had such walls, as did most textile mills in New England built after 1880 (Fig. 5-2). Nevertheless, the deep piers obstructed sunlight and had other unwelcome effects: they either projected into a building so the interior wall could not be flat, or projected out on the facade, which created a buttressed look that the architect may not want.

Some designers reduced the bulk of the piers by reinforcing them with iron columns and thereby dispensing with some of the masonry. The building usually cited as the first skeleton building—the Home Insurance Building in Chicago (1883–85)—actually had an iron-reinforced wall in

Fig. 5-2. Elevation of a bearing wall (1882) with the masonry concentrated in piers (also called *pilasters*). Boston Manufacturers Mutual Fire Insurance Company, *Special Report No. 10* (Boston, April 1, 1882).

its street facade (the rear wall was load-bearing) (Fig. 5-3). Architects soon eliminated the masonry altogether by placing loads entirely on the wall columns. Skeleton walls first appeared in the back sides of buildings, around light-courts, in the 1880s, while the street fronts and party walls of these buildings were of ordinary load-bearing construction. The real pioneer skeleton frame buildings were those with skeleton frames in their street facades: such buildings demonstrated the feasibility of using the frame to support all the loads.[18]

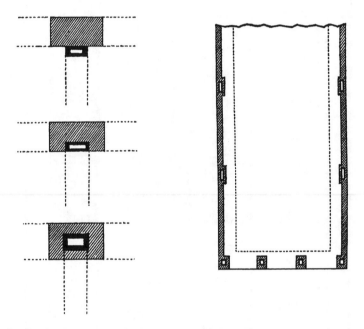

Fig. 5-3. In the drawing on the left, a contemporary architect, William J. Fryer, illustrates the evolution of the wall column. Designers first placed iron columns near brick piers, to strengthen them (*top*), then placed the columns inside the pier (*middle*), and eventually dispensed with the pier and used the column alone for support, although the column continued to be enclosed for fire and weather protection. The plan on the right compares wall dimensions for a tall building with bearing walls and one with a skeleton frame: the dotted line shows the amount of the floor area that bearing walls would cover, and the hatched lines, the amount covered using skeleton frame construction. William J. Fryer Jr., "Skeleton Construction, A New Method of Constructing High Buildings," *The Architectural Record* 1 (1891–92).

Because many architects contributed to its development, most contemporary writers agreed that no single person should get credit for creating the skeleton frame. According to the ironmaker, architect, and building official, William J. Fryer Jr., the skeleton system "may be said to have been incubated, rather than invented, and the simple, triumphant method of constructing the most marvelous of modern buildings is found upon examination to be but an enlarged use of preceding methods."[19] It might be expected that the skeleton frame would have developed most rapidly in New York City, the center of fireproof construction, but this was not the case. Although the skeleton system appeared almost simultaneously in Chicago and New York City, structural innovation proceeded at a faster pace in Chicago in the 1890s. For this reason, as a New York structural engineer acknowledged in an early textbook about the new system, "the majority of writers call it the 'Chicago Steel Skeleton Construction.'"[20] In neither city was the system introduced specifically to build structures taller than they could be built with current practices; rather, the system solved some of the problems inherent in tall, wall-bearing buildings. The early skeleton buildings were no higher than other wall-bearing buildings of the day.

At first, two different systems were sometimes called skeleton construction. In both cases, the building's frame carried the floors and roof; but in one, the frame also carried the "walls" (the outer envelope), whereas in the other, the walls supported themselves. Some contemporaries referred to this latter system as "cage construction" and reserved the term *skeleton construction* for the former.[21] The walls in a cage construction building had to be thicker at the bottom because they carried their own weight, and so took up more space in the lower floors than did walls in true skeleton buildings. Still, they were thinner than load-bearing walls, and the wall columns in a cage building could be lighter than those in a skeleton building, since they did not have to carry the weight of the envelope. In some cases, the savings in construction costs from using lighter wall columns may have outweighed the disadvantage of losing some interior space.

These kinds of cost and performance trade-offs lay behind a designer's decision to adopt one system or another in the early days of skyscrapers. The pioneer skyscrapers, typically eight to twelve stories, often contained a mix of framing systems. A designer might put a skeleton wall on the street facade, which allowed him to install large windows, and build load-bearing party walls. In New York City's first skeleton frame building, the Tower Building, only the lower stories had skeleton walls; the upper stories had bearing walls. The skeleton system was developed to solve specific problems; designers seemed unconcerned about keeping the structure pure.

A different set of conditions spurred the development of the skeleton frame in Chicago and New York City. In Lower Manhattan, a situation that magnified the disadvantages of tall, wall-bearing structures was the small size of the typical building lot. Rather than combining lots, which sold for princely sums, owners ordered tall buildings for tiny parcels. What is believed to be the pioneer skeleton frame building there, the 129-foot Tower Building (1888–89), was erected on a lot that measured just twenty-one-and-a-half feet wide. If the owner had put up a conventional building that also complied with the city's building law, its party walls at the base would have been thirty-two inches thick, leaving a space about sixteen feet across on the inside. In order to increase the rentable area, the building's architect, Bradford Gilbert, proposed carrying the party walls on girders that spanned between wall columns. Since this did not conform with the building code, he first had to convince the city's Board of Review to grant an exemption. Gilbert argued that the lower floors were actually part of the building's foundation and therefore not subject to the rules on wall thickness. Although skeptical, the board permitted him to proceed. He made a skeleton frame of cast iron columns and rolled iron girders and finished the floors with hollow tile from Maurer & Son.[22] People worried that the skinny building would overturn, but Gilbert signaled his confidence by taking the top floor for his own office.

Designers, builders, and materials manufacturers in New York immediately recognized the significance of Gilbert's structural innovation, and a second skeleton frame building started going up while the Tower Building was still under construction. This second one, the Lancashire Fire Insurance Company Building (1889–90), was another tall (ten-story) building on a narrow lot (twenty-four feet, two inches wide). By 1891, the skeleton frame had a chronicler, William J. Fryer Jr., who described the evolution of the system in the first volume of a new architecture periodical, *The Architectural Record.*[23]

For Chicago developers, the problem the skeleton frame solved was not narrow lots, but the difficulty of building secure foundations in the city's compressible soil. By the latter part of the nineteenth century, grade level for land in the Loop, Chicago's commercial center, was about fifteen feet above water level. At water level, the ground was a layer of stiff clay called "hardpan"; between this and grade level was soft material—fill and sand.

The next level of stiff soil lay about 65 feet below the hardpan, and bedrock, fully 80 to 120 feet below. Since they lacked practical means to make deep foundations, designers instead devised methods of putting in foundations that did not break through the hardpan.[24] In a wall-bearing building, the more stories, the greater the weight on this fragile, hardpan base; the skeleton frame offered a means to reduce weight while increasing height.

Although usually considered the first skeleton building in Chicago, the thirteen-story Tacoma Building (1887–89) contained a mix of structural systems: skeleton walls along its two street facades and load-bearing walls at the side, the back, and through the interior. The idea of the skeleton walls at the Tacoma may have come from Sanford Loring, the terra cotta manufacturer and architect, who proposed making the walls out of his hollow blocks and placing them on the building's frame. Perhaps the first Chicago building to have a full skeleton frame was the sixteen-story Manhattan Building (1889–91). Other early examples include the Siegel, Cooper & Co. and Ludington Buildings (both c. 1891).[25]

Although owners liked the thin skin of skeleton frame buildings, the system had its critics, of which the New York architect-engineer George B. Post was the most vocal. Post designed some of the earliest skyscrapers, but in the 1890s he began to oppose tall buildings generally and skeleton frame buildings in particular, because he considered them structurally unreliable. He warned that the skimpy covering around most wall columns—four to twelve inches of masonry—did not protect them adequately against fire and water. From his observations of the effects of moisture on steel, he concluded that over time it would corrode more readily than iron, and therefore he preferred the cage system, with its substantial walls that better protected the frame. Post pushed the cage system to its practical limits in his thirteen-story World (Pulitzer) Building (1889–90) in New York City, which, counting the six-story dome, rose 275 feet above the street. The walls carried only themselves, but still were over eleven feet thick at ground level. A knowledgeable contemporary described the building as an "extreme example" of cage construction.[26] As might be supposed about a building with such thick walls, the World stood on what was for Lower Manhattan a large piece of real estate. But owners of small lots wanted tall structures, too; for them, skeleton construction was the only practical choice.

Another important advantage of skeleton construction, and to a lesser degree cage construction, was speed. Historically, fireproof buildings took longer than average to finish. Wall-bearing buildings, whether fireproof or not, had to be completed in a sequence, from the bottom to the top. In a skeleton or cage building, once ironworkers erected the frame, work could commence on any floor, or on several floors simultaneously. Reduced construction time meant big savings for a developer—shortening the term of his construction loan and speeding the day when the building could be rented or sold. Because skeleton frame buildings could go up quickly, owners used the system for low buildings as well as high ones.

The early skeleton buildings contained the same materials as the wall-bearing, fireproof buildings with one exception, and this was steel. Even though some called it the "steel skeleton" system, steel was by no means an essential ingredient for skeleton buildings. The skeleton frames of many pioneer skyscrapers—the Tower, Tacoma, and even the tallest building in its day, the sixteen-story Manhattan Building—contained cast and wrought iron columns and girders. Around 1890, designers began to use steel beams instead of wrought iron; a few years later, they adopted steel columns. Nevertheless, many architects continued to specify cast iron columns in skeleton frame buildings through the early 1900s.

The products of Henry Bessemer's converter and later methods of refining pig iron mechanically came to be called "steel," but these metals bore little resemblance to the ancient material called steel that was used in cutlery and tools. Some wondered why the new metal was called steel. William Starrett, the prominent builder, considered it to be "a wholly new iron, intermediate between wrought and cast."[27] Later known as "mild steel," this new material contained less carbon than cast iron but more than wrought iron. Carbon is the component of cast iron that makes it strong in compression but brittle; the virtual absence of carbon in wrought iron is what makes it malleable and strong in tension. Mild steel, the "intermediate" product, combined the advantages of both materials: it was as strong as cast iron in compression and stronger than wrought iron in tension, making it suitable for columns as well as beams.

Two processes for manufacturing the new steel—Bessemer and open-hearth—were available by the second half of the 1860s when Americans first began to make mild steel, but initially all American plants were the Bessemer type. The number of Bessemer plants grew in the 1870s and steel output increased steadily, even as its price slumped during the depression of that decade. The mills that rolled railroad rails absorbed practically all of this raw steel; indeed, the rail mills often built steel mills in order to be assured a supply of steel. At this time, engineers did not consider rail-grade steel reliable enough for structural shapes, and they avoided it. This made no difference to the steelmakers, since they could sell all they made to the rail mills. They had no reason to produce anything better.[28] Thus, despite the growing domestic steel industry, architects and civil engineers specified wrought iron for structural members. Before 1885, very few American bridges and no buildings contained mild steel members.

Yet mild steel had qualities, in addition to its strength, that made it attractive to constructors. One of these was "homogeneity." Manufacturing the new steel involved melting pig iron completely and tapping off impurities; the resulting product was uniform throughout. Moreover, a Bessemer converter could produce a large quantity of steel in one blow. Structural shapes could be rolled from one uniform mass rather than from piles of wrought iron welded together. Steel could be riveted, an advantage in tall skeleton buildings with relatively small bases that were liable to twist and bend in strong wind. To resist wind pressure, such buildings had to have stiff, braced frames, which was only possible with rivets. Connections to cast iron members, which could not be riveted, were made with bolts, and this sort of connection often left some play in the joint. The hot riveted connections between wrought iron and steel members, in contrast, joined members into essentially one continuous piece. Finally, constructors preferred rolled to cast metal because manufacturing imperfections were more common in the latter and hard to detect.

However, constructors knew little about the how mild steel would perform in structural applications. They needed the sort of information that could only be gained through impartial tests, but these would be expensive and difficult to conduct. Beginning in the early 1870s, architects and engineers pressed the federal government to sponsor such tests. They made the same arguments that the previous generation had made when it sought information on the new rolled iron beams: the test results would allow the government to use materials more efficiently in its fireproof buildings and thereby save tax dollars.[29]

Congress agreed and in 1875 authorized funds "to determine by actual tests the strength and value of all kinds of iron, steel, and other metals . . . and to prepare tables which will exhibit the strength and value of said materials for constructive purposes."[30] A public-private group with the unwieldy name United States Board Appointed to Test Iron, Steel and Other Metals developed a schedule of tests and set up a testing laboratory at the Army Arsenal in Watertown, Massachusetts. In its short life, the board accomplished little more than bringing into being a unique testing machine, designed and built by the brothers Albert and Charles Emery. Unlike earlier testing machines, which could handle samples of material or scale models of structural members only, the huge Emery machine could test full-sized members. It could measure, with "wonderful" accuracy, specimens as long as thirty feet and as fine as horsehair.[31] Practically all of the board's appropriation went to purchasing this machine and in 1879, when it ran out of money, the board disbanded. Many questions persisted about steel, including the central ones concerning its reliability and durability.

Constructors also avoided steel because it cost more than cast and wrought iron. High cost had long discouraged builders from using iron beams, and steel beams were even more expensive than iron. But the stubbornly high price of beams was largely the result of price fixing by the manufacturers through their cartel, the Beam Association. Only a handful of American mills rolled beams; the seven members of the Association, by one contemporary estimate, had roughly an 80 percent market share among them.[32] American mills had tried to control prices from the very start of beam manufacture. In the 1850s, with Phoenix Iron and Trenton Iron the sole domestic producers of beams, Trenton's owners tried to come to an "agreement" with Phoenix because, "The business is as yet too inconsiderable for active competition."[33] Mills that entered the business later preferred to join rather than compete with Phoenix and Trenton, and they formed what critics called the "mill pool." The difficulties of setting up a beam mill aided this collusion; as the trade periodical *The Iron Age* observed, "it requires a working capital out of all ordinary proportion to the tonnage of product, and . . . it takes long and costly experience to become successful in the business."[34] Moreover, some of the beam makers—Phoenix Iron, and Carnegie, Kloman & Company and its successors—established subsidiary bridge building companies and therefore had built-in customers for their products. Foreign imports posed no serious threat, mainly because of the high tariffs on imported iron. Mill owners lobbied hard to keep the tariffs high, which made imported beams uneconomic outside the seaport cities. In 1888, AABN found that beams brought in from Belgium could be had at the ports, even with the cost of freight, insurance, and an estimated 120 percent ad valorem tariff added in, for less than the price of American beams.[35] Why, AABN wondered, should beams cost so much more than rails? "There is no reason," AABN concluded, "why steel rails should be sold at our mills for a little more than a cent a pound while steel beams, rolled in the same way, out of the same material, should cost more than twice as much."[36] These extortionate prices were not merely unfair, according to AABN, but they had serious social consequences, since they discouraged owners from constructing fireproof buildings (Fig. 5-4).

But by the mid-1880s, output from the ever expanding steel industry outstripped the needs of the railroads, and steelmakers had to search for new customers. Just at this point, Carnegie-Phipps & Company offered to substitute steel beams for wrought iron ones to fill out its contract for the Home Insurance Building in Chicago. William Jenney, the building's architect, accepted the proposal and completed the upper floors with steel beams—the first to be used in an American building.[37] Jenney became a strong advocate of steel rather than iron beams. Indeed, Chicago architects generally preferred steel beams: during the very busy building season of 1890, rolling mills could not keep up with their demand for steel beams for "fireproof work." Despite the delays that resulted, Jenney urged architects to stick with steel because "there is no other detail in an architect's practice by which he can save so much money for his client."[38] Since steel was stronger than iron—a fourth to a third stronger, according to Jenney—a

Fig. 5-4. An 1888 cartoon illustrating the influence of industrial cartels in Congress; the "steel beam trust" is the most prominent moneybag. "Bosses of the Senate," *Puck* 24 (1888).

designer could use smaller steel sections to carry the same load, and this effectively eliminated the price difference between the two. On the downside, Bessemer steel, the principal kind at this time, still had a reputation
for unreliability, and indeed, Jenney advised architects to put an agent at
the mill to test the tensile strength of each blow and inspect all beams for
defects. Even though such supervision added to a project's cost, Jenney still
felt the economics favored steel. He expected beam prices to fall and "before long every hotel, apartment house, theater, and schoolhouse will be
fireproof."[39]

Jenney's prediction about prices proved correct, and in the 1890s, the
prices of both iron and steel beams began a downward trend. In 1892, the
despised mill pool—that "combination of manufacturers, which has kept
up the price of iron beams . . . for more than twenty years"—dissolved.[40]
By the middle of 1892, steel beams that in 1891 had sold for about 3½ cents
per pound could be had at Pittsburgh and Chicago for 2 cents per pound. In
addition, their quality improved, especially as rolling mills began to use
the more reliable open-hearth steel for structural shapes. This conjunction
of lower prices and an improved product cut the effective price of structural metal to half of what it had been two years before. Beam imports
ceased. All this boded well for fireproof construction. The editors of
AABN, echoing Jenney, foresaw a new era, since, "with the lower prices for
iron, its field of use will be greatly extended. [With iron at two cents a
pound] many a building will bring in rents sufficient to pay interest on the
cost of fireproof construction . . . , which would not do so with iron at three
cents; and fireproof construction will be adopted, to the great advantage of
all concerned."[41]

Switching from iron to steel beams posed little difficulty for designers:
whether in iron or steel, the sections came in the same shapes. Columns
were another matter. Before 1890, the only kind of wrought iron column
used by designers to any extent in buildings was the Phoenix column,
manufactured by the Phoenix Iron Company. It consisted of a tube formed
by joining together three or more flanged sections along their length; the
size of the column varied with the number of sections. For additional
strength, the mill could place iron plates between the sections, which increased the outer dimension and stiffened the column (Fig. 5-5). This column served as a key element in the wrought iron truss bridges manufactured by the Phoenix Bridge Company, a subsidiary of the Phoenix Iron
Company, and probably the bulk of Phoenix Iron's columns went into
bridges. Yet Samuel J. Reeves, the column's inventor and Phoenix Iron's

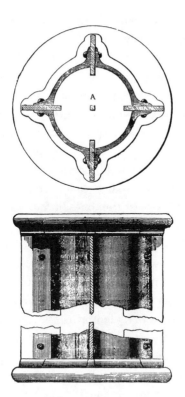

Fig. 5-5. Early Phoenix column. Section through a column (*top*) shows the flanges riveted together with iron plates at the junctures for reinforcement. Elevation (*bottom*) shows the cast iron capital and base for the columns. The column illustrated, used in Samuel White's circa 1867 store in Philadelphia, was filled out, presumably with plaster, to create a smooth cylinder. *Architectural Review and American Builders' Journal* 1 (August 1868).

vice president, also intended it for buildings, as an alternative to cast iron columns.[42] He patented the column shortly after a notorious accident: the collapse of a large textile mill in Lawrence, Massachusetts, caused partly by the defective cast iron columns used in its construction. The mill's builder did not know the columns had defects; and indeed, flaws such as cavities in the metal, stress lines from unequal rates of cooling, and walls of unequal thickness could not be easily discovered. Moreover, the practical length of cast iron columns was limited by the difficulty of making very long ones. Casting long columns required pouring iron into both ends of a mold; should the metal chill and fail to unite in the center, the column would be weak at a structurally critical place. Nevertheless, even with flaws, cast iron columns rarely failed.[43] But they already had an undeserved reputation for being unreliable, and the Lawrence mill accident only made it worse. Phoenix columns, made with iron shaped by rolling and hammering, were less susceptible to such defects.

Architects used Phoenix columns to a greater extent in the second half of the 1880s, but most still opted for less costly cast iron. Moreover, apart from avoiding defects, designers had no strong reason to adopt wrought iron columns. The subject of wind bracing had not yet received much attention. Until the middle of that decade, architects employed wrought iron and steel columns like cast iron ones, in post and beam fashion, with beams resting on top of the columns. Not until 1894, in the reconstruction of the Reliance Building in Chicago, did a designer introduce angle brackets to attach girders to columns, a form of connection necessary to create a continuous, truly rigid frame.[44] But not all skeleton buildings required this sort of frame. The factors that influenced the choice of columns included a building's height relative to its footprint, the resistance provided by the walls, and whether its tenants used machinery that caused vibration.

Another factor that determined whether a building would have steel or cast iron columns was whether a bridge engineer worked on its structural design. American bridge engineers loathed cast iron columns and invariably specified steel columns. They had long since abandoned cast iron for compression members in bridge trusses, and in the 1890s, switched from wrought iron to steel for these members. Chicago architects tended to employ civil engineers more often than did their counterparts in New York City, and consequently, steel columns showed up more frequently in Chicago's tall buildings than in New York's. The Rand McNally Building in Chicago (completed 1890), with a frame designed by the engineering firm of Wade & Purdy, was one of the first to have steel columns and the first to use columns in the Z-bar shape popular with bridge engineers. Wade & Purdy also specified steel columns for the Caxton building in Chicago, completed in the same year.[45]

The price of steel columns remained high while steel beam prices fell, but eventually column prices came down, along with the rest of the nation's economy, in a depression that started in 1893. In 1894, construction began on the first building in New York City with an all-steel frame—the American Surety—and from then on, the number of buildings with such frames increased.[46] By the latter part of this decade, wrought iron structural shapes had vanished from the mills' catalogues. "Our product," the

1896 edition of Carnegie Steel Company's *Pocket Companion* explained, "is exclusively steel."[47] Steel columns came in many styles, in addition to the Z-bar type, including ones fabricated from plates, U-channels, angles, and small bars, and special patented shapes like the Phoenix column and its imitators (Fig. 5-6). The architectural iron industry suffered as a result of this trend. Of the many foundries in New York City that once made structural castings, for example, only about a half dozen remained at the close of the century.[48]

Nevertheless, although not flourishing, the foundry business did not disappear. Steel columns may have been cheaper than formerly, but they still cost more than cast iron columns. Moreover, some architects assumed that steel would behave like wrought iron, which rusted more readily than cast iron, and in a fire, weakened at a lower temperature and expanded faster than cast iron, and so was more likely to wrench or push out the walls. For these reasons, and undoubtedly also because they were accustomed to using them, many architects continued to specify cast iron columns and even cast iron spanning members for the walls of skeleton frame buildings, where they could be exposed to moisture and fire. New York City architects in particular continued to use cast iron columns: at the turn of the century, perhaps three-quarters of all new buildings of twelve stories or less contained cast iron columns.[49]

In fact, no one knew how well any of the materials of the "modern" iron and steel-framed, hollow tile buildings would perform in a fire, because few had been tested. Would a very hot fire break the tile fireproofing, expose the metal structure, cause it to expand or twist, and lead to the destruction of the building? In the early 1890s, fires in several new, tall buildings provided some reassurance. An 1891 fire in the Lumber Exchange Building in Minneapolis tested two types of systems: the semifireproof and iron and hollow tile systems. Built in two phases, the older nine-story section had iron columns and girders and wood joist floors; both the frame and floors were protected with tile. At the time of the fire, the owner was erecting an eleven-story, iron and hollow tile addition and also adding two floors, built of iron and tile, on top of the old building. In order to raise a water tank up to the new roof on the old building, the contractor cut a hole through all the floors; he had not yet closed up the holes, which left the wood floor frame exposed. A fire started in a neighboring building and broke into the Lumber Exchange, igniting the exposed wood around the holes, and burned up to the roof. The heat from the fire softened the metal legs of the water tank atop the hole, which then fell through, strik-

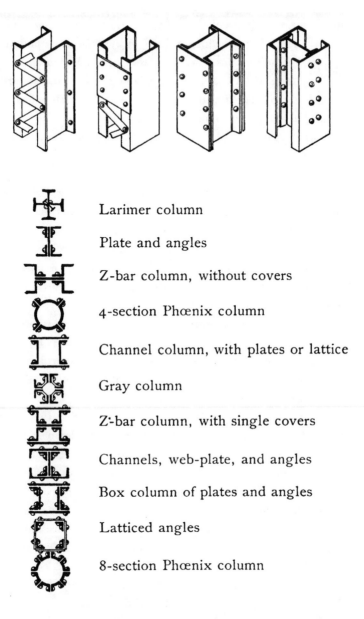

Larimer column

Plate and angles

Z-bar column, without covers

4-section Phœnix column

Channel column, with plates or lattice

Gray column

Z-bar column, with single covers

Channels, web-plate, and angles

Box column of plates and angles

Latticed angles

8-section Phœnix column

Fig. 5-6. Various forms of steel columns. Top shows isometrics of typical steel columns made of channels connected by plates, lattice, or I-sections. These nonpatented columns, including the Z-bar, were the forms in which steel was most widely used at the turn of the century. Below are other steel column sections showing some of the less commonly used patented and restricted forms—the Larimer, Phoenix, and Gray columns. J. K. Freitag, *Architectural Engineering.*

ing beams and dislodging fireproofing tiles as it crashed to the ground. To compound the trouble, the fire department fought the blaze ineffectively. In the end, all nine of the old stories had been gutted, but remarkably, the two new top floors remained intact. The new wing suffered little damage: the hollow tile in this part did not even crack. But since no tenants had moved in yet, the building contained little combustible material on the inside and so escaped a severe test.[50]

The following year, a serious fire in the nine-story Chicago Athletic Club finally provided experts with "an opportunity to judge the actual efficacy of the steel and fireproof construction."[51] Still under construction when it caught fire, the Club was a cage-style building with a steel frame and porous tile floor arches, partitions, and column coverings. The fire started on a Sunday morning and spread to several floors before being discovered. Finished with wood paneling and wainscoting, and with piles of wood scaffolding, construction refuse, and painters' materials lying around, the building contained much fuel: the fire became so hot it fused the glass in several windows. Despite the inferno conditions, the Club's fireproofing held up on most floors, and the insurance adjusters estimated the cost of repairing the steel frame (as opposed to the finishes and much of its stone facade, which were ruined) to be minimal. Experts agreed that by neglecting fireproofing details, such as putting wood nailing strips between the fireproofing tiles, the building's architect allowed the fire to do much more structural damage than it might have. Rather, as two prominent engineers who studied the fire's aftermath concluded, "the metal parts of a building, if thoroughly protected by fireproofing properly put on, will safely withstand any ordinary conflagration."[52] Following repairs, the Club opened the next year. The lessons of this fire: mind the fireproofing details and restrict the amount of combustible finishes and construction materials inside. While these two fires seemed to vindicate the new fireproof construction methods, they did not help designers choose among the great range of fireproofing products that came on the market in the 1890s.

New Fireproofing Products
in the 1890s

The expanding demand for fireproof building materials in the 1890s gave a tremendous boost to the fireproofing products industry. New companies entered the business; manufacturers introduced new materials along with improved versions of old products. The increased competition drove down prices. Some observers welcomed these developments, arguing that "cheap and effective methods" created by the "fireproofing interests" encouraged the spread of fireproof buildings and made cities safer. Others countered that the intense competition forced manufacturers to drop their standards and flood the market with "cheap and nasty" products, which resulted in inferior buildings.[53] Of the new products introduced in this decade, the two most important were a form of hollow tile floor arch called "end-construction" and concrete floors.

While a variety of patterns for flat arches existed before the 1890s, all were of a type called "side-construction," meaning that the cavity of the block ran parallel with the beam. Another style, in which the cavity of the block ran perpendicular to the beam, was produced in England, and a few Americans patented such arches in the 1870s, although no one manufactured them. Structural theory predicted that these "end-pressure" or "end-construction" arches would be stronger than the side-construction type, since the walls of the blocks, running at right angles to the beams, would function like beams themselves. In the 1880s, a flat end-construction arch was tested in Chicago, and by the close of the decade, arches of this type had been installed in several buildings in Chicago and elsewhere in the West[54] (Fig. 5-7).

In December 1890, the relative merits of two types of arches were put to a widely publicized test. The architects of the fireproof Equitable Building in Denver received bids from three contractors for hollow tile floor arches, one of whom, Thomas A. Lee, claimed that while his price was highest, his product was both stronger and more fire-resistant than those of the other bidders. He had proposed an end-construction style of floor made of a relatively new material, porous tile, rather than the usual solid fire clay. Lee asked that the products be put to a trial, and the building's architects consented, provided that the other bidders, Pioneer Fireproof Construction Company and Wight Fire-proofing Company, would participate. They did, and all agreed to specifications for the test. At the time, no standards for testing fireproofing materials existed. Whenever a manufacturer wanted to demonstrate the merits of his product, he subjected it to a test of his own devising, and potential customers then had to decide how well the test conditions coincided with the ones the product might have to withstand in their buildings. Moreover, because manufacturers did not use identical tests, consumers could not readily compare results. In the Denver case, probably for the first time, three different floor products were subjected to identical tests, so at least these floor systems could be compared

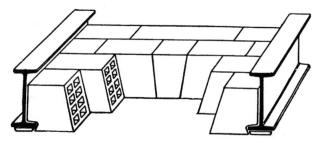

End-construction Terra-cotta Arch.

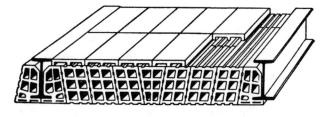

Side-construction Terra-cotta Arch. Bevelled Joints.

Fig. 5-7. End-construction (*top*) and side-construction (*bottom*) arches. In the newer, end-construction style, the webs of the blocks run perpendicular to the beam. In the side-construction style, the webs run parallel with the beam. On the side-construction model illustrated here, small tile blocks on top of the arch fill the space level to the upper flange of the beams. J. K. Freitag, *Architectural Engineering.*

with each other. The tests were a load test; a shock test (in which a log was dropped on the floor arch); a fire and water test; and a continuous fire test. The sample floors were to be tested to destruction, a harsh and not very realistic standard.

In the end, it was no contest: Lee's arch outperformed the others decisively. It sustained two to three times the loads that broke the other floors, and ten drops of the log, which shattered the other floors on the first blow. Moreover, the porous terra cotta in Lee's arch withstood the fire and water and continuous fire tests better; the solid tile in the other two floors disintegrated over the course of the test.[55] Lee won the contract. Peter Wight, the fireproof construction expert and one of the unsuccessful bidders, suggested that trial had been unfair because Lee's tiles were made especially for the test while he and Pioneer supplied blocks from their regular stock. Nevertheless, Lee's sample was exactly what he proposed to furnish. Wight offered another, more trenchant, objection to the test: the floors had not been evaluated in light of any specific performance requirements, but only to find out which system provided the "most." Had there been performance specifications, Wight's floor, although not as strong Lee's, nevertheless might have been strong enough for the purposes of the building. "When all the service that is called for is obtained from a material," he observed, "and it cannot be made lighter, nothing is gained by introducing stronger systems of construction. We do not build of granite because it is stronger than brick when brick will answer, and if the end-pressure system should be more expensive than the old one, there is every reason why we should adhere to the old one."[56]

His argument changed few minds. Instead, hollow tile manufacturers quickly turned out their own versions of end-construction arches. Pioneer introduced a large, multicelled, end-construction type block, which could create spans of up to twelve feet. Henry Maurer & Son offered an end-construction arch like Pioneer's, called the "Excelsior" floor. Illinois Terra-Cotta Lumber Company also offered end-construction-style arch patterns. Experts began to recommend end-construction arches generally, and the type became the standard. Chicago architects used end-construction arches early on, and after about 1894, according to the engineer Edward Shankland, who worked on many tall buildings, they used the system almost exclusively. Likewise in Philadelphia, by the end of the century, all tile floor work was end-construction.[57]

In addition to developing end-construction arch styles, tile manufacturers continued to try to make the blocks simpler. In the more elaborate styles of side-construction arches, such as those with radiating webs, every unit in the arch had a different shape, which made them expensive to manufacture and troublesome to lay. Manufacturers developed floor patterns that contained only three different block shapes—the skewback, key, and intermediate blocks—and then even simpler systems, consisting of two different blocks or even only one. An example of a single block floor was the "Herculean," introduced by the Maurer firm at the end of the decade. The floor could achieve very long spans—up to nearly twenty-three feet, depending on the depth of the block—and therefore could cover rooms by resting on partition walls and so reduce the number of girders.[58] The Herculean floor system was used in a variety of buildings at the turn of the century, including schools, factories, libraries, and apartment houses (Fig. 5-8).

"Herculean" Flat Arch
(Patented May, 1898, and February, 1900)

This construction will at once appeal to Engineers and Architects as presenting an up-to-date fireproof method
Section showing method

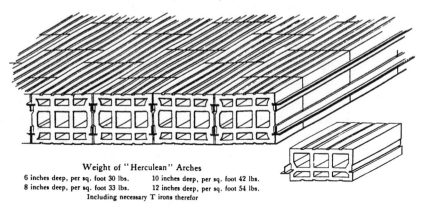

Weight of "Herculean" Arches

6 inches deep, per sq. foot 30 lbs.   10 inches deep, per sq. foot 42 lbs.
8 inches deep, per sq. foot 33 lbs.   12 inches deep, per sq. foot 54 lbs.
Including necessary T irons therefor

Fig. 5-8. Herculean floor arch, manufactured by Henry Maurer & Son, made of uniform blocks. Henry Maurer & Son, "Illustrated and Descriptive Catalogue of the 'Herculean' Flat Arch (Terra-Cotta) and 'Phoenix' Hollow Wall Construction," 1908.

The structural tile industry had come a long way from its early days in the 1870s. At the turn of the century, hollow tile was the most widely used material for floors, partitions, and roofs in fireproof buildings. The 1900 U.S. Census of Manufacturers found that the nation's output of clay fire-proofing products—the category for "hollow brick, fireproofing, and terra-cotta lumber"—had quadrupled over the decade. Companies in this line of work could be found in many states: the largest share of the output, by far, came from New Jersey, with Ohio and Pennsylvania following. The National Fire-proofing Company of New York and Chicago, reportedly the "largest company devoted to the manufacture and erection of hollow-tile fire-proofing-material," had taken over several competitors, including one of the oldest, the Pioneer Company. Old companies still doing a large business were Henry Maurer & Son, the Haydenville Mining and Manufacturing Company (Ohio), and Illinois Terra-Cotta Lumber Company (Chicago). While the inspiration for hollow tile floor arches came from England and France, Americans made the system their own. It became as much a part of the 1890s skyscrapers as steel. Meanwhile, Europeans did not develop their tile floor systems or even use them to a large extent.[59]

Reinforced concrete, another system of construction pioneered by Europeans, was the other important fireproof construction material developed in the 1890s. Americans used a substance they called "concrete" in the early nineteenth century, although its composition differed from that of modern concrete. The term *concrete* was applied to any mixture that included a bonding material and "aggregate," typically sand and larger masses, such as broken stone, furnace ash, or even rubble from demolished buildings. The bonding material might be a "natural" one, such as plaster, lime, or cement produced directly from stones, or "artificial" cement, commonly known as "Portland" cement, which was made by combining various natural ingredients according to a recipe. Thus, through the nineteenth century, concrete was not a uniform material. Not until the 1890s did the term come to mean mainly (but still not exclusively) a material made with Portland cement.

Plain concrete is weak in tension although strong in compression, and around the last quarter of the nineteenth century, Americans began to combine it with steel or iron, which provides tensile strength, to make a versatile structural material. Around the turn of the century, Americans commonly called this material "concrete steel." Concrete interests pushed for another name, "reinforced concrete," and this more descriptive name stuck.

Americans used concrete very little in the first three-quarters of the nineteenth century. Military engineers were among the first to employ it, in the foundations and walls of fortifications. Civil engineers later used it to build dams, canals, and bridge supports. American architects had little experience with the material. Some used it to level up brick arches, and Chicago architects made a kind of foundation, called a grillage foundation, of iron rails and concrete, adapted for the city's compressible soil. Several companies made concrete building blocks—also called "artificial stone"— and builders could buy presses to make their own blocks by the 1860s. The Fire-Proof Building Company used concrete to make some of its hollow blocks. However, no one used concrete in monolithic, load-bearing spanning applications.[60] The only sorts of concrete spanning systems used in the United States before the 1890s were Gilbert's and Cornell's concrete-filled, corrugated iron decks.

In England and Paris, in contrast, architects adopted slab concrete fireproof floors around the mid-1800s. Concrete floors had a number of potential advantages over brick arches, principally lightness and greater span. In the 1850s, a concrete and iron joist floor called the Fox and Barrett floor became popular in Britain. Parisian builders used similar sorts of floors, principally the "Système Vaux" and "Système Thausné," the latter much used in apartment blocks put up during the redevelopment of Paris under Baron Haussmann. In the 1860s, British inventors introduced a variety of concrete floor systems, the most popular types being Phillips's, Dennett's, and W. B. Wilkinson's floors.[61]

American architects knew about these systems, but did not try to replicate them, mainly because of the scarcity and high cost of a key ingredient: "hydraulic" cement or lime, so named because of its ability to set under water. By the mid-nineteenth century, several British manufacturers could produce a strong, hydraulic material similar to modern Portland cement. The Paris region contained an abundance of a natural cement and plaster that became impervious when it set. Americans manufactured natural hydraulic cements, but these had the drawbacks of setting rapidly and not being as strong as Portland cement. Before the 1870s, Americans produced little if any Portland cement and for some time after domestic production began, most Portland cement used in the United States came from abroad.[62] This made it very costly, except in one part of the country, which surprisingly was California. Faced with a shortage of traditional building materials, California builders began to use concrete building blocks in the 1850s. The market for this product in the West induced Ernest Ransome, son of the English cement manufacturer Frederick Ransome, to relocate to San Francisco in 1870 and set up a business to make artificial stone using his father's patented process.[63] Although it had much farther to travel, at this time, English Portland cement cost less in California than in the East. One explanation for the price advantage, offered by an unnamed contemporary "expert," was that it went West as ballast: "English Portland cement . . . was taken to California, freight free, as ballast for sailing ships that had taken out California wheat."[64]

One of the first efforts to make a building out of monolithic concrete was made by William Ward, a prosperous manufacturer. He built a fireproof home for himself in Port Chester, New York (1873–77), out of "béton combined with iron"—*béton* being the French word for concrete. Ward conducted his own tests to determine the best combination of materials with which to make concrete, as well as the sizes of iron beams and bars to be used as reinforcement. The project required "four thousand barrels of imported Portland cement," and no one at this time followed his example.[65] Ward's experience highlights another reason why Americans did not take up concrete floors: little was known about how to build reliable floors. Only an enthusiast like Ward would go to the trouble and expense of experimenting with iron and imported cement.

There was another such enthusiast, the maverick American inventor Thaddeus Hyatt, and in 1877, he published the first work that explained how concrete and iron worked together. In the 1870s, Hyatt began a search for "cheaper and more reliable fireproof constructions than those in common use," which involved experiments on concrete and iron floors conducted at David Kirkaldy's testing laboratory in London. Several features of the concrete floor systems then in use in Europe troubled him. He believed the great weight of concrete in the slab type floors overloaded the joists, and that the lower flanges of the beams in these floors were not adequately protected from fire. He also suspected that iron and concrete expanded at unequal rates and therefore in a fire, the expanding iron could break up the concrete slab. Finally, he wished to devise a system that required fewer expensive beams. From his tests, Hyatt discovered that the flanges of iron beams added no strength to a concrete floor and indeed, a grid of light iron placed in the lower part of a spanning element—a concrete floor slab or a beam—provided sufficient tensile strength. Hyatt also tested the fire resistance of his reinforced slab by placing a sample of it on top of a furnace and starting a fire. He found, contrary to his expectation, that concrete and iron expanded at similar rates, and he therefore judged the concrete and iron system he created to be fireproof. Hyatt's report on his findings, "An Account of Some Experiments with Portland-Cement-Concrete Combined with Iron, as a Building Material," became a seminal work, and the following year he patented a reinforced concrete system.[66]

In California, the engineer P. H. Jackson, another concrete construction pioneer, began to put Hyatt's ideas into practice soon after they appeared; he held the West Coast license to Hyatt's patents.[67] In the 1880s, Ransome also expanded into concrete slab construction, at first to improve his firm's illuminating sidewalk panels. Ransome made these panels—a framework with glass blocks designed to cover cellars underneath sidewalks—out of concrete with reinforcing bars, along the lines that Hyatt recommended. But Ransome added a twist, literally: rather than using straight, smooth bars, he used twisted ones, reasoning that they would stay more securely in the concrete. This mechanical connection, by keeping the reinforcing bars in place, allowed him to dispense with having to make an iron frame in tension. He patented his idea for rough bars in 1884. Specially shaped bars became features of many patented concrete systems introduced when concrete construction took off in the early twentieth century.[68]

The earliest concrete floors appeared around 1890 in the East, but these were mainly one-off installations. For example, in 1889, the architect Frank Furness patented a concrete floor similar to Fox and Barrett's, but Furness's floor called for economical wood and iron fabricated joists rather than rolled I-beams. Furness installed this floor in the fireproof library he designed for the University of Pennsylvania (1888–90). Other architects put concrete floors in Philadelphia-area buildings around this time, including the Wilson Brothers, who rebuilt the interior of the American Philosophical Society with concrete fireproof floors in 1891. The floors consisted of iron beams filled between with concrete reinforced with small iron bars.[69]

Concrete construction entered a commercial phase in the 1890s, coinciding with a sharp upturn in American production of Portland cement and a drop in its price. A big upswing in cement output occurred around 1895–96, when Americans abandoned European methods of manufacture and adopted ones better suited to local circumstances. Whereas in 1890 the American Portland cement industry turned out about 336,000 barrels, in 1895 it produced three times this amount and the following year, nearly five times as much. This growth resulted from an increase in the number of cement plants as well as greater output per plant. By 1897, the number of barrels produced domestically for the first time exceeded the number imported; imports fell off slowly while domestic output rose dramatically, reflecting the huge demand for cement. The price for a barrel of American Portland cement fell from $3 in 1880 to about $2 in 1890, and then to a little over $1 in 1900.[70]

Two different sorts of concrete construction developed. The earliest kind was concrete floors for metal-framed buildings, like those in Philadelphia—essentially American versions of the floor systems that had been used for decades in Britain and Paris. The later type included buildings made largely or entirely of reinforced concrete, with a concrete frame as well as concrete floors, of which the William Ward house was an early example. This kind of construction—the wholly concrete building—was a new thing: few reinforced concrete buildings had been erected anywhere in the world before the 1890s.

In the early 1890s, construction companies began to propose concrete floors for skeleton buildings. They offered several types of floor systems, including both plain (unreinforced) concrete arches and reinforced types. Builders used similar methods to construct both kinds of floors. First, they erected a form between the beams or walls to support the concrete until it set. For a plain concrete floor with a wide span, the form would be arched, to create a floor that supported loads through compression. If the floor was reinforced, builders usually put up a flat wooden platform, then placed reinforcing metal between the beams and covered it with concrete. Concrete floors could be finished like hollow tile floors, that is, with wood nailing strips laid on top to hold the finish floor (Fig. 5-9).

Concrete floors had several advantages over hollow tile. First, they could be thinner and therefore lighter. They filled in irregularly shaped spaces better than hollow tile and created a smooth surface on top, without the breaks or joints that opened should the tiles settle unevenly. Concrete floors could also be made exceptionally strong. Trade and engineering magazines published photographs of concrete floors piled high with bars of pig iron or sacks of cement. On the downside, concrete floors required a great deal of water at the job site; they took longer to dry than tile floors

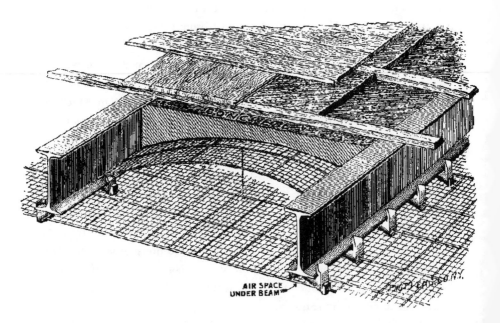

Fig. 5-9. Roebling Concrete Arch, one of the first kinds of concrete floors to be widely used. Metal lath forms a permanent center between the beams; it was filled with concrete, which was not reinforced. The floor was finished underneath with a wire lath and plaster ceiling, suspended under the floor, and finished on top with a wood floor. *The Architectural Record* 7 (1897–98).

and so could not be walked on as soon; and workers had to be closely supervised to make sure they made the concrete properly. Some constructors found the lightness of concrete floors a disadvantage: one of these, Edward Shankland, the Chicago engineer, preferred deeper tile floors because they provided additional lateral stability.[71] But concrete's biggest advantage after the price of cement fell was its comparatively low cost. Concrete floors could be built by unskilled workers while hollow tile work required skilled masons, a difference concrete advocates stressed, and the reason union masons initially opposed the system.[72]

Hollow tile manufacturers and their supporters tried to drive out the upstart concrete contractors as soon as they entered the business. Concrete's impressive load-bearing capacity could not be denied and its cost advantage made it a competitive threat. On one vital matter, however, hollow tile seemed to have the upper hand, and this was in its fire resistance. Until the end of the 1890s, little was known about how concrete would perform in a fire. Fire tests of small briquettes of cement and mortar had showed that concrete, although it might hold together after being subjected to high heat, could be permanently weakened.[73] Tile advocates latched on to this concern as their best weapon against concrete. For example, an article in *The Architectural Record*, written to counter the "aggressive advertising" of the concrete builders, claimed "exhaustive tests" had proven that concrete could not survive even an ordinary fire. The writer charged the concrete interests with spreading false information based on "a specious fire-test conducted under the auspices of the New York Building Department." At the same time, the writer reported favorable results of fire tests on hollow tile floors, although these had been conducted by the same, presumably unreliable, building department.[74] By aggressive advertising, he probably had in mind an article published in the magazine a few months before that mentioned the successful performance of John A. Roebling's Sons Company's concrete floors in an 1897 test. The Roebling Company also advertised in *The Architectural Record*.[75]

These New York tests, in fact, came about in order to resolve the debate between the concrete and the tile factions, which fought for contracts in the nation's biggest market for fireproofing products, New York City. Only the materials specifically listed in the city's building code could be used in fireproof buildings, and this list did not include concrete. Owners could petition the city for an exception, to use concrete in a particular project. But concrete floor manufacturers felt this process treated them unfairly: they wanted their products, which they considered equal to tile, to be allowed as of right. To try to settle the question of how well concrete floors compared with tile, in 1896 New York's building superintendent arranged for tests, on an unprecedented scale, of the fire resistance of various fireproof floor systems. The systems tested included both monolithic concrete and combination tile-concrete floor systems, as well as several kinds of wooden floors—fourteen types in all. The concrete systems included an unreinforced Roebling arch; a Thomson concrete slab, similar to the Fox and Barrett floor; a Columbian Fireproofing Company floor, made of special cross-shaped steel reinforcing bars in a concrete slab; and a Bailey floor, which consisted of metal decking, with dovetail-shaped corrugations, covered with concrete. Four other concrete floor systems—Clinton Wirecloth, Manhattan, Expanded Metal Fireproof Construction Company, and Metropolitan—were variations of a type: a concrete slab reinforced with wire fabric or cables. No two concrete systems used the same formula for concrete; some formulas called for plaster or natural cement rather than Portland. This was the first time a variety of systems had been subjected to the same test, which would yield comparable results. It was also a major undertaking. The Building Department found safe sites and designed a standard test structure: a small brick-walled kiln. Each company built its own floor system over the top of the structure so it formed the roof of the kiln. Building Department officials supervised the installation, watching to see that the floors put in were like those the companies installed in actual practice.[76]

The department arranged for very severe tests, which involved burning the floors for about five hours, much of the time at temperatures over 2,000 degrees F; afterward the Fire Department turned hoses on the floors for fifteen minutes. The floors carried normal loads (one hundred and fifty pounds per square foot on the central arch) during the fire tests, and then, after cooling, were loaded with six hundred pounds per square foot, which remained in place for forty-eight hours. The concrete floors performed well: although damaged to some extent—more by the Fire Department's water than by fire—all remained intact. Following more tests conducted by the manufacturers at the department's direction, the building superintendent decided to allow concrete floors. But the victory for concrete was short-lived: the city administration changed and the new building chief reinstated the former rule.[77] This reversal outraged civil engineers, not to mention the concrete floor contractors. They lobbied a committee then working on a revision to the city's building code to get their products included as fireproof materials, and engineers and the periodical *Engineer-*

ing News supported this cause. *Engineering News* was a frequent critic of
the city's Board of Examiners, the body that ruled on applications to use
concrete, regarding it as technically unqualified and backward. It considered the claims for the superiority of tile over concrete to be unfounded.

The pro-concrete side received unsolicited support from a conflagration
in Pittsburgh in May 1897, which burned several new fireproof buildings,
one of which, the Methodist Book Publishing Building, had concrete slab
floors. Two others, the Horne store and Horne office building, had hollow
tile floors—hard tile in side-construction in the former, porous tile in end-
construction in the latter. The fire, therefore, provided a field test of the
fire resistance of concrete and its performance relative to tile. Architectural and engineering periodicals covered the fire thoroughly, and their
conclusions covered the gamut; commentators seemed to see what their bi-
ases predisposed them to find in this complicated event.

On one point, no one disagreed: the structure of the Methodist Building
survived largely unscathed even though its contents had been consumed in
the fire. While this would seem to be an endorsement for concrete, many
observers remained unconvinced. Corydon T. Purdy, the respected structural engineer, expressed the views of those in the tile and steel camp
when he argued that this incident did not provide a fair test. In a paper he
read to the American Society of Civil Engineers, he suggested that the
Methodist Building survived as well as it did because it escaped the intense
heat that the tile-floored buildings suffered. In the discussion that followed his talk, engineers who had inspected the Pittsburgh buildings for
insurance adjusters voiced the pro-concrete view. In their opinion, the
Methodist *had* been exposed to intense heat; moreover, its performance
was consistent with the findings of the New York City Building Department's tests. They concluded that "properly made concrete was as good if
not better than burnt clay."[78]

The contest between the tile and concrete camps over New York City's
fireproof floor business took a bizarre turn in 1899, when a state legislative
committee established to investigate government corruption in the city added floor construction to its agenda. Called the Mazet Committee after its
chairman, Robert Mazet, the Republicans that dominated it formed it for
the purpose of embarrassing the city's Democratic boss, Richard Croker.
At least, this was how contemporaries regarded the "investigation." According to AABN, the hearings were "a device of one set of political cor-
ruptionists to extort, by a pretense of exposure, from another set of cor-
ruptionists, a larger share than it now enjoys in the plunder of the great

city of New York."[79] In addition to investigating the Police Department and other agencies, the committee tried to smoke out the conspiracy in the Building Department to foist concrete floors on unsuspecting New Yorkers.

The main target of the committee's Building Department hearings was the Roebling Construction Company. Formed at the end of the 1890s, the company carried on the fireproof construction work of its predecessor, the John A. Roebling's Sons Company. As it happened, one of the investors in the new company was Frank Croker, Richard Croker's son. Moreover, the elder Croker's nephew represented the Roebling Company in petitions before the Building Department.[80] The Republican members of the committee apparently hoped to show that the Building Department granted exceptions, and changed specifications for city buildings, in order to benefit a company in which Croker, or at any rate his son, had a financial interest. Indeed, the Building Department did permit builders to use concrete on a case by case basis and it also allowed concrete to be used in the city's own buildings. But since these decisions benefited many concrete firms, and not only the Roebling Construction Company, they did not prove favoritism. Thus, to make a case for corruption in the Building Department, or at least incompetence, the committee's interrogator tried to prove that concrete floors were inherently defective and should not be allowed at all. The spectacle of attorneys, in what was supposed to be an investigation of official corruption, grilling witnesses about the structural properties of concrete must have been comical. This line of attack soon fizzled: the committee could not show that concrete was a dangerous material just because Democratic officials permitted builders to use it. Nevertheless, the right to build concrete floors unconditionally in New York City awaited a 1903 revision to the building law.

Around the same time, complete reinforced concrete buildings—with concrete frames as well as concrete floors—began to go up on a commercial scale in the United States. In 1904, a new periodical devoted to encouraging owners to build with concrete appeared, *The Cement Age*. Edited by the president of the Cement Manufacturers Association, the magazine reprinted articles from engineering magazines, published technical monographs, and catalogued the growing number of concrete buildings. The concrete buildings erected during the first decade of the twentieth century were mainly factories and warehouses—the sorts of buildings that owners rarely built fireproof and, since they were not tall, building codes did not require to be fireproof. Thus, while concrete floors had a cost advantage over tile floors, which encouraged builders to use concrete, concrete buildings had a cost disadvantage relative to the alternatives, which would have been slow-burning or ordinary construction. The main reason owners paid the premium and built in concrete was because they wanted fire-resistant buildings.

Fire Underwriters and
the Fireproof Building

A persistent theme among fire safety advocates was that fire insurance companies had the power—in the form of the rates they charged—to make buildings safer. All owners of commercial property and business tenants carried fire insurance: no investor would take a mortgage on an uninsured building; no manufacturer or merchant would extend credit on uninsured goods. Fire insurance greased the wheels of commerce, as the Continental Fire Insurance Company's president, Francis C. Moore, explained; with fire insurance, "ventures are made, without hesitation, which would appall those embarking in them if liable to miscarry through a single fire."[81] For many architects and building owners, the necessity of insurance to businessmen gave the companies leverage over their policyholders. They argued that the companies should use this power to discriminate against unsafe buildings—penalize them with high rates and reward the good ones with low rates—and thereby rid cities of firetraps. Indeed, those who advocated this sort of "insurance control" assumed that owners would save enough in fire insurance costs to cover the added expense of putting up a safe building in the first place. In short, safety would be free. They also believed that it was in the underwriters' self-interest to adopt a rate structure that had these effects, because fire loss would decline.

But the fire insurance industry did not see things this way. It was not their business to prevent fires. On the contrary, from the perspective of the traditional underwriter, "Fires were what made business for underwriters." They had no obligation to reduce the nation's fire waste. Rather, "it was the function of insurance simply to distribute the fire-loss, and if people preferred to burn their property, it was not [the underwriter's] business to interfere; it was his business to see that plenty of premiums were collected to pay the losses." This attitude infuriated the industry's critics, who charged that the industry actually made cities more hazardous, by insuring unsafe buildings at low rates and charging safe ones too much, to make up the difference. Insurance companies got away with this, they believed, because the public did not understand insurance matters.[82]

Of course, fire insurance companies did set rates according to the risk a building represented, to some extent. But fire safety advocates wanted the companies to more finely discriminate among buildings and not deviate from the rates they set, by, for example, giving a break to the owner of a hazardous building in order to earn a high premium. Unfortunately, the intense competition in the fire insurance industry made it difficult for companies to stick to a rate schedule, even if they wanted to. They were more likely to charge whatever it took to win or keep a customer. Moreover, coming up with a detailed schedule that priced the innumerable characteristics of a building that contributed to its safety or hazard was a devilishly difficult task. How much did one feature or another add to the likelihood of a fire starting? What amount of losses might be expected from buildings with such a feature and therefore what was the proper charge for having it? To make this sort of calculation, companies needed a considerable amount of data, which they simply did not have and, more to the point, were reluctant to go to the trouble to collect.

To add to the obstacles, no one company had a sufficiently large number of customers of a particular type to set rates for the type based solely on the company's own loss experience. For example, it might happen that during one year, none of the stables a company insured suffered a fire. The company would be wrong to conclude from this that stables are unlikely to have fires and therefore should be charged very little for insurance; the following year might be very different. To set correct rates, the company would have to know the long-term loss experience of stables overall, in other words, know the average loss for stables as a group over a number of years. But no such general information about properties existed. To create it, many companies would have to pool data on their loss experiences, and this was something the stock fire insurance companies refused to do. They considered their company's loss experience a trade secret and were not interested in turning such information over to their competitors. If they had a profitable class—say, stables that had not burned—they would not want other companies to know about it. Moreover, in order to group like properties, the companies would have to make uniform reports; companies did not necessarily collect the kind of information that would be required to classify properties. Finally, even if companies did classify properties in a uniform way and faithfully collect and report the details of their customers' properties and losses to some organization, the difficulty of tabulating these data for thousands of properties, in the days before computers, was probably insurmountable. And what organization would receive, analyze, and prepare reports on these data? One group that tried to fill the role was the Committee on Statistics of the National Board of Fire Underwriters, the trade association for the industry. It called on members to contribute data so that it might compile statistics that could be used to set rates, but companies would not oblige.[83]

Thus, insurance companies set rates based on their own internal data, "judgment," and, of course, competitive pressures. They usually lumped properties into groups according to occupancy (office, theater, hotel, etc.) and for buildings within a group, categorized roughly according to their construction (i.e., masonry or wood walls) and then set a rate for the class. Underwriters might add charges to the class rate for what they considered dangerous features of a building, but did not make deductions for good features. This class rating had the advantage of being inexpensive to administer since it did not require companies to make a thorough inspection of each property. But the approach had many critics even within the industry. One early insurance textbook urged underwriters to "place each risk on its own merits" (risk being the industry term for property to be insured). He went on to condemn class rating, "that is, to charge the same, or about the same rate, on all buildings constructed of the same material (brick, for instance)," a practice he found "as absurd as it is general."[84] In 1874, the president of the National Board of Fire Underwriters, who belonged to the reform wing of the industry, urged his colleagues to adopt a more detailed system of rating, called "schedule rating." By continuing to use class rating, and offering the same rates for "risks very diverse in their physical hazard," insurance companies could not "influence the erection of safe and substantial buildings."[85]

Apart from the extra work involved in implementing such a system, many underwriters worried the approach might work too well and cut their income from premiums. Both insurance agents—the independent salesmen who worked on commission and received a percentage of the premium amount as payment—and the companies themselves benefited from high premiums. Neither had much incentive to discover ways to reduce rates on good buildings, or to raise them of bad risks and thereby drive away the customers who made hefty payments. Agents had no financial stake in the losses the companies they represented might suffer. Fire insurance companies likewise valued customers that paid high premiums: the income that companies earned by investing premiums was important to their profitability.

Nevertheless, traditional underwriters dreaded conflagrations, since they imposed large losses all at once, and building up reserves to cover

such disasters was difficult. If companies raised rates in order to build a reserve, their customers might complain that rates were excessive, and some
competing company, less concerned about long-term survival, would try to
take their customers by offering lower rates. To reduce competition, large
insurance companies supported state insurance laws that excluded under-
capitalized companies from doing business in the state. Underwriters' organizations lobbied communities to improve their fire-fighting services
and to adopt building regulations aimed at reducing the likelihood of conflagration. Industry representatives served on building code-writing committees.

With the number of tall buildings increasing in many cities, a regulation that underwriters particularly sought was restrictions on building
heights and volumes. They worried that fires in high buildings, beyond
the reach of firefighters' hose streams, would spread uncontrolled along
surrounding roofs and perhaps engulf a district. In the 1890s, localities for
the first time imposed limits on building heights. A little more than a year
after Boston's terrible Thanksgiving Day fire, Massachusetts passed what
was probably the nation's first height-limitation law. This 1891 law capped
building heights in all Massachusetts cities, except Boston, at 125 feet. Boston, which had its own separate building law, followed suit and the following year restricted heights to 2½ times the width of the street, up to a
maximum of 125 feet. At the same time, it made more buildings subject to
the fireproof construction rules, by lowering the threshold for fireproof
buildings to 70 feet from 80; required large hotels as well as theaters to be
fireproof; and restricted the floor area of nonfireproof buildings to 10,000
square feet. The following year, the city tightened the volume provision,
reducing the allowable undivided area of nonfireproof buildings to 8,000
square feet.[86]

Likewise, the Chicago City Council, after years of debating height limitations, passed a height limitation ordinance in 1892. Although the mayor
vetoed this bill, the city did enact a height limitation law the following
year, which capped buildings at 130 feet.[87] Many cities adopted height limits during the 1890s and first decade of the twentieth century. Only New
York City, home of the greatest concentration of tall buildings, bucked the
height restriction trend, until 1916, when it adopted a new zoning law that
called for facade setbacks.[88] In most cities, these height "limits" were not
especially restrictive; nevertheless, fire safety was an important rationale
for imposing them.[89]

While fire underwriters pressured local governments to impose building regulations, state governments in turn pressured the insurance industry to set rates more fairly. Beginning with Ohio in 1885, state legislatures began to pass laws that forbade insurance companies or agents from joining into boards of underwriters for the purpose of fixing, maintaining, or controlling insurance rates. By 1891, eight states had enacted these "anti-compact" laws or anti-trust laws that also applied to insurance companies, and bills had been introduced in several other states.[90] The insurance industry strenuously opposed such measures, arguing that the anti-compact laws actually made it more difficult for companies to set fair rates. Companies argued that unless they could meet in local insurance associations to share information, they could not establish "fair, conservative rates, based on large experience."[91] But the public suspected the trade groups met for the less savory purpose of fixing rates rather than to share data in order to determine fair ones.

In the early 1890s, while the industry worked to fend off adverse regulation, its reform wing renewed efforts to establish and put into effect "schedule" rating, partly to demonstrate to the public that rates were rational and therefore fair. Led by Continental Insurance Company's president Francis C. Moore, a committee of underwriters prepared a comprehensive schedule rating system for city buildings. In 1893, it unveiled an elaborate schedule called the "Universal Mercantile Schedule," which tried to list every factor that "an ideal underwriter . . . would take into account in fixing a rate." An important goal was to establish a system that could be applied anywhere in the United States—a necessary feature if the companies were to quell complaints of disparate treatment of some cities compared to others. The schedule also allowed rate concessions for buildings that were better than the norm. Setting a high standard for the baseline building and giving breaks when a building exceeded this, the committee believed, would eventually encourage owners to erect safer buildings.[92] The committee published its schedule to inform owners in advance how various features of a building would be rated. In the early twentieth century, several insurance theorists introduced alternative schedules.

The schedule rating movement was an important advance in the process of rationalizing rate setting, and in this sense benefited the stock fire industry, but it did not stop lawmakers from introducing legislation that the insurance industry opposed or from launching investigations into industry practices. People both in and outside the industry praised schedule rating and believed it encouraged owners to improve their buildings. The stricter local building laws large cities were adopting at this time also worked in this direction. Arguably, the new rating schedules, especially the more favorable treatment eventually given to fireproof buildings, helped tip the scales in favor of fireproof construction, even thought the savings in insurance could not completely cover the added cost.

Triumph: Fireproof Skyscrapers
Survive Conflagrations

The terrible fire that destroyed Baltimore's downtown in 1904 provided the most thorough test of the new fireproofing methods to date. Early in the morning on February 7, a fire in the basement of a wholesale dry-goods store swept up an elevator shaft and ignited products stored on the upper floors, causing a series of explosions. Firemen arrived quickly at the scene, but already flames had engulfed the surrounding buildings. A gale-force wind blew the fire from its starting place, on the west side of the downtown, to the northeast, when a wall of substantial public buildings (the fireproof courthouse, post office, and city hall) stopped its advance. Then the wind's direction changed and turned the fire east, toward the banking and wholesale part of the city. Overwhelmed, Baltimore's fire department called for reinforcements, and companies hurried in from Philadelphia, New York, Washington, Wilmington, and York, Pennsylvania. But the firefighters struggled in vain against the spreading fire. In desperation, the mayor ordered buildings in the path to be blown up, to create fire-breaks, but the explosions only fueled the fire. With the wind blowing southeasterly, the firefighters made a stand on the east at a natural barrier, a stream called Jones Falls, and helped by water thrown from the harbor by the powerful fire tug, the *Cataract*, they held back the fire. On the south, the fire stopped at the harbor, and by the afternoon of the following day, it finally burned itself out. In the end, it destroyed about 2,500 buildings on roughly 150 acres. One fireman was killed when a wall collapsed on him and many others were injured, but no civilian died.[93]

This was not the first American urban conflagration in the new century: that dubious distinction went to Jacksonville, Florida, which suffered an extensive fire in 1901. During the winter of the following year, another fire destroyed over five hundred buildings in Paterson, New Jersey. But unlike jerrybuilt Jacksonville or central Paterson, with its mix of combustible business blocks and residences, the burned district in Baltimore contained many substantial, brick buildings including perhaps a dozen modern-style fireproof buildings. Among the latter were the twelve-story tile and steel

Calvert Building, built by the renowned contractor, George A. Fuller Company, and the sixteen-story tile and steel Continental Trust Building, designed by the famous Chicago architectural firm, D. H. Burnham & Company. Although a decade older than these two, the ten-story fireproof Equitable Building (completed 1892) was a massive specimen of fireproof construction, with thick bearing walls, cast iron columns, and floors of segmental, hollow tile arches.[94] Fire experts, architects, civil engineers, and builders converged on Baltimore to study the consequences of the fire and especially the performance of the fireproof buildings. What did it show about the modern methods of structural fire protection?

To the layperson, the utter devastation permitted no judgment other than that "fireproof" construction was a fraud. Fire underwriters joined in the censure. The Baltimore fire, one asserted, "shows that modern fireproof buildings will not even retard a great conflagration"; another warned owners not to expect any concessions for fireproof construction and to stop complaining about rates.[95] Some fire protection experts, notably those in the factory mutual field, also took the opportunity to criticize "so-called" fireproof construction. Edward Atkinson got back at his critics, asserting "the first lesson taught by this conflagration is for the underwriter to expunge the term 'fire-proof' from his dictionary."[96]

In contrast, architects, engineers, and engineering writers who specialized in fireproof construction concluded that the fire vindicated modern fireproofing methods.[97] The fire tested the fireproof buildings severely: it was an inferno by the time it reached the fireproof buildings. Consequently, they had been exposed to exceptionally high temperatures and for a long time. Investigators found fused glass, brass, and even cast iron in the wreckage, which indicated temperatures of around 1600 to 1800 degrees F.[98] Despite this assault, the walls of Baltimore's "fireproofs" remained standing and formed barriers to the spread of flames, whereas the ordinary buildings ended up as heaps of rubble (Fig. 5-10).

The experts attributed the damage suffered by the fireproof buildings to faulty details and shoddy construction in some cases, and to a general lack of protection against exposure fire: specifically, inadequate coverings for windows. Nevertheless, in nearly all cases, the frames of these buildings came through intact and the floors survived, although in poor condition. The interior of the Equitable Building sustained the most damage, but this resulted from heavy safes on upper stories falling after the fire weakened the floors.[99] Such defects could be corrected with existing methods and materials. The solution to the fire problem, therefore, was not to

Fig. 5-10. Baltimore after the 1904 conflagration, with fireproof buildings standing. Mississippi Wire Glass Company, *Reconnaissance of the Baltimore and Rochester Fire Districts*... (c. 1904).

go back to the old "brick and mortar" system, as some laymen urged, but to build *more* fireproof buildings. An *Engineering News* editorial made this point:

> The experience of Baltimore has illustrated afresh what every well-informed engineer already grasps, but what the general public is slow to grasp, that no human ingenuity can devise a building of reasonable cost and utility which will withstand unimpaired the attacks of such a conflagration as raged last week in Baltimore. On the other hand, if all buildings in the business section of a city were reasonably fire-resistant, as they might be at moderate cost, large areas could not be fire-swept and utterly destroyed.[100]

In other words, fire-resistive buildings could help avert conflagration, but not if they stood as islands in sea of firetraps. Cities would be secure when a sufficient share of all their buildings were fireproof: "fireproof structures can only stand in fireproof cities."[101]

On a matter that continued to divide the fire protection experts—the fire resistance of concrete in a conflagration—the Baltimore fire shed little new light. Only a few small buildings in Baltimore had concrete floors, and the National Fire Protection Association, in its report on the fire, concluded that all of them had escaped intense heat. Other experts—those who favored concrete—disagreed, asserting that the buildings with concrete floors had been exposed to severe conditions and had come through well. This was the conclusion of Charles Ladd Norton, head of the Factory Mutuals' Insurance Engineering Experiment Station: he made two visits to Baltimore shortly after the fire and found that in two buildings with reinforced concrete floors, temperatures had reached the point of softening cast iron. Yet the floors showed no serious damage, a finding consistent with his experimental results on the fire-resisting properties of concrete. Likewise, Carl Greishaber, an engineer with a large New York architectural firm, studied the ruins of Baltimore's fireproof buildings and concluded that concrete floors withstood the fire better than tile.[102] But since so few buildings in the burned area had concrete floors or frames, their performance could not resolve this question definitively.

For many in the building world, definitive information arrived two years later, in April 1906, from the fire that followed the great San Francisco earthquake. The San Francisco conflagration grew from small blazes that began in several locations after the earthquake—in some cases, started by people trying to cook food in houses with broken chimneys. With the city's water mains broken by the earthquake and the fire department leaderless after its chief was killed when a weakened chimney fell on him, the fire took over the city. For three days, it burned its way through four square miles—the business center as well as wooden residential areas—and killed hundreds of people. The amount of property destroyed, valued at $350 million, made this by far the worst urban conflagration in U.S. history and among the world's worst fires ever. Today, many Americans know about the 1906 earthquake, but at the time, the engrossing event was the relentless, all-consuming fire.

Even without the earthquake to set it off, San Francisco was ripe for a sweeping fire. An estimated 90 percent of the city's buildings were wood frame, and many of these stood among the large, new hotels, office buildings, and stores in the city center.[103] The National Board of Fire Underwriters' Committee of Twenty surveyed San Francisco in 1905 and compiled a long list of hazardous conditions. These included buildings with large, unobstructed floor areas; numerous tall buildings; frequent unprotected openings and light-wells; and a lack of protective devices and equipment, such as shutters, wire glass windows, sprinkler systems, or private wells, tanks, or pumps. The report warned that these factors, combined with the many narrow streets and hills in the city that could slow down firefighters, created ideal conditions for a conflagration, and concluded famously, "San Francisco has violated all underwriting traditions and precedent by not burning up."[104] Yet San Francisco also contained many fire-resisting buildings in its downtown: over fifty could be considered fire-resisting, about forty-five of which were fireproof and the rest modeled on slow-burning construction. Despite the danger of earthquake and fire in the city, San Francisco had a number of skyscrapers, which possessed the most up-to-date steel frames, including the fifteen-story Call/Claus Spreckels Building and the fifteen-story annex to the Chronicle Building.

San Francisco's fireproof buildings differed from those of other U.S. cities in that many had concrete floors, the result of the Bay Area's early start in concrete construction. About three-fifths of the fifty or so fire-resisting buildings had concrete floors and a few of these had concrete frames. Many had Roebling floors—the first sort of concrete floor system used widely in San Francisco buildings—including such ten-plus-story buildings as the Kohl, Merchants Exchange, Mutual Savings Bank, and Call/Claus Spreckels Buildings. Others had floors made of concrete reinforced with expanded metal. Several contained both tile and concrete, most commonly tile column fireproofing and tile partitions. A few contained both concrete and tile floors, such as the Call/Claus Spreckels Building, which had concrete floors up to the seventh and hollow tile floors above. Thus, the nation's worst conflagration, quite coincidentally, occurred in the only city with enough structural concrete to provide many examples of the effects of fire on concrete. Still, the test was not a perfect one, because few buildings suffered from the fire alone; they also were damaged by the earthquake and by dynamite, thanks to the futile efforts of an amateur dynamite squad.[105] Moreover, San Francisco's buildings had a reputation for shoddy workmanship. These factors complicated the task of sorting out the effects of the fire from damage arising from other causes.

Inevitably, experts drew divergent conclusions about the performance of concrete and the relative merits of concrete and tile fireproofing. Of the numerous studies of San Francisco's fireproof buildings in the fire, perhaps the most comprehensive were three reports prepared for the U.S. Geological Survey (USGS) and published together as a study on the effects of the earthquake and fire on structures. Each was written by an engineer: Richard Humphrey, head of the structural materials division of the USGS; Captain John S. Sewell, Army Corps of Engineers; and Frank Soulé, dean of the college of civil engineering of the University of California, who was on the scene. While they agreed on some matters, on the critical one, "the old and vexed question of concrete vs. terra-cotta construction," their opinions differed.[106] Richard Humphrey concluded that concrete withstood the fire better than tile, which cracked and spalled in the heat and did not insulate structural steel adequately. He concluded: "Solid concrete floors proved satisfactory, though concrete in San Francisco was of a very poor quality, and flimsy concrete stiffened with light metal passed as reenforced concrete." He went on: "If reenforced concrete of the [poor] quality described could give such satisfactory results in meeting the extraordinary condition of the San Francisco earthquake and fire, it is evident that much greater satisfaction would have been given by the use of first-class material."[107]

Captain Sewell was less impressed with concrete's performance, and he judged tile and concrete about equal, although he criticized the typical methods used to build both. In the end, he came down on the side of well-made hollow tile, since he believed that concrete, although appearing uninjured, inevitably dehydrated in a fire and would be cracked internally. For this reason, he recommended that concrete floors be insulated underneath with metal lath and plaster ceilings, to protect them. Frank Soulé, being familiar with local building conditions, concluded that concrete fireproofing had worked well, but it was heavy, and in the Bay Area, good cinders for making cinder-concrete were scarce. Moreover, trade unions opposed concrete and for this reason, only one all-concrete building existed in San Francisco. In his comprehensive textbook on fire protection, the concrete skeptic Joseph K. Freitag reported these diverse views and declined to make an unequivocal statement about concrete's comparative merits.[108]

While the experts debated, the market made up its mind, and it voted for concrete. There was enough evidence that concrete performed as well as tile, and it cost less. Henry Ericsson, the Chicago builder, recalled that in the first decade of the twentieth century, reinforced concrete was "coming in" to Chicago. Although initially reluctant to use it, architects soon were "swept along" by the "tremendous driving power" of the cement and steel manufacturers, "resulting in the use of reinforced concrete even where it was of doubtful economy or value."[109] Likewise in Philadelphia, many of the new fireproof buildings in the early twentieth century contained one of the numerous concrete floor systems then available. Ninety percent of the postconflagration buildings in Baltimore had concrete floors or column protection, or both. In the decade following the San Francisco fire, even John Sewell became a convert, describing concrete as "highly efficient" in its fire-resistive qualities.[110]

Thus, with the spread of concrete construction in the first decade of the twentieth century, all the materials necessary for making a fire-resistive building had been introduced. As experts constantly emphasized, these materials could be used imperfectly, in ways that defeated their capabilities. Buildings called "fireproof" could be nothing of the kind, and many builders still thought fireproof construction meant little more than avoiding combustible structural materials. As a textbook on fire protection admonished, "the mere fireproofing of steel work, putting in this or that material in floor construction, does not constitute a heavenly dispensation to construct every other part of the building in a way which is no better than that used in the veriest firetrap."[111] However, if used properly, in a building planned so as to isolate horizontal from vertical spaces; with structural elements protected against heat and water; windows and doors properly guarded against exposure fires; and with internal protective equipment such as standpipes and sprinklers, then a building could be made effectively fireproof. With building laws compelling owners to build them, fireproof buildings would spread. Fire protection experts worked to see that cities adopted good laws and that the laws were competently and energetically enforced.

The debates over limiting building heights did not end when cities enacted height caps. Real estate developers periodically lobbied to raise the bar. People who opposed tall buildings continued to argue that skyscrapers were fire hazards. Yet building codes everywhere required tall buildings to be fireproof, and experience showed that buildings could be fireproof. Many developers asserted that these tall, fireproof buildings made cities safer because they would serve as barriers to the spread of conflagration. While fire protection experts agreed with this argument—indeed, they used it when urging cities to require fireproof construction for a greater variety of buildings—they did not agree that this granted a license to developers to put up any number of structures as high as they wanted.

As a practical matter, few buildings conformed to the highest standard. In a statement to New York City's 1913 commission investigating the merits of regulating building heights, Ira H. Woolson and F. J. Stewart, both engineers representing fire insurance interests, explained that most high buildings in New York could be rated "only as of *fairly efficient fire resistive construction.*" With the exception of warehouses and factories, tall buildings rarely had automatic sprinklers, and owners declined to put in wired glass and shutters to protect against fires on the outside. These conditions, combined with other common structural defects prevalent in various kinds of large buildings, made them vulnerable not only to fires spreading to them from outside, but also weakened their ability to resist the spread of a fire on the inside. Woolson and Stewart warned, "The possibilities of a fearful holocaust in the burning of such a building is apparent."[112] Therefore, they urged height limits as a failsafe plan, to compensate for the danger posed by a concentration of less-than-perfect fireproof high-rise buildings.

6

CALAMITY

*Slow Acceptance of
Adequate Fire Exits*

There is no "fire-proof" building that cannot be made death-proof as well.

EDWARD CROKER, New York City Fire Chief, 1911

Before 1911, two problems dominated the work of fire safety experts: fire prevention on the one hand—measures to avoid unwanted fires—and fire protection on the other—designing buildings so as to control and confine fires. Fireproof construction is an example of a fire protection measure: a means of putting together a building so that it would sustain little damage in a fire and thereby reduce property loss.

That the purpose of a fireproof building was to safeguard property is borne out by contemporary definitions in technical literature. Frank Kidder's widely used *Architect's and Builder's Pocket-Book* defined a fireproof building as one made and finished entirely of noncombustible materials, in which all the structural parts capable of being injured by heat or water were protected. By the end of the nineteenth century, experts added the stipulation that for a building to be considered fireproof, vertical spaces on the inside had to be isolated from horizontal spaces. Thus, the structural engineer Corydon T. Purdy defined a fireproof building as one that, in addition to being noncombustible, would "confine an internal fire to any room in which it occurs, without material injury to the rest of the structure." A broader definition gained acceptance in the early twentieth century, which added internal fire suppression to the features of the best sort of fireproof buildings. At this time Joseph Freitag's comprehensive fire protection handbook came out, in which the author defined fire protection to include, "control of fire through construction, control by means of first aid or departmental work, detection of fire by automatic means, and safety against exposure hazard."[1] What is striking about these definitions, to the modern mind, is the absence of any mention of the safety of a building's occupants. In no case did safeguarding human life figure in the objectives of fireproof construction.

Although they did not articulate safety to life as a goal, the experts nevertheless believed that by making buildings fireproof, they were at the same time making them safe for occupants. When floors did not burn through and the structure remained intact in a fire, occupants had more time to escape or to be rescued. A noncombustible stairway, shielded by masonry walls and entered through fire-resistive doors, would survive longer in a blaze and be less likely to fill with smoke and flames than an ordinary one. Automatic sprinklers could extinguish a fire, or at least keep it in check and douse the smoke. Buildings that possessed these features were safer for occupants than ordinary ones.

Yet the experts focused exclusively on perfecting the fire resistance of structures and thereby overlooked the fact that fireproof buildings could still contain hazards for occupants. If nothing else, the building's contents could burn. Therefore, a sufficient amount of safe exits or places of refuge were necessary even in fireproof buildings. But the fire protection fraternity did not see the necessity. They considered emergency egress a problem for buildings that were *not* fireproof and therefore out of the realm of their concerns. Thus, in the nineteenth century, the men with the most knowledge about fires, and how to design, construct, and manage buildings to protect them from fire damage, simply did not apply themselves to the issue of how to evacuate burning buildings. Like so many lessons in the field of fire safety, the experts learned the error of equating structural safety with the safety of occupants the hard way. They underestimated how quickly the smoke, heat, and poisonous gases from burning contents could kill people. It took several deadly fires in fireproof buildings in the early twentieth century to make this point. Once they understood the problem, they broadened their work and made egress a central focus.

While the fireproof construction experts did not deal with emergency egress in the nineteenth century, other individuals and organizations did, principally housing reformers, architects, social welfare groups, and the agencies that enforced building and labor laws. Through the efforts of these reformers, by the turn of the century, all large cities as well as many states had enacted laws that required redundant or protected exits in certain kinds of hazardous buildings, to allow occupants to escape in a fire. Yet for the most part, these laws required only modest, often ineffective, measures. Like building rules generally, advances in egress regulations tended to track catastrophic events, in this case, individual building fires that killed many people.

The History of Egress Laws

Before the 1850s, many American cities and towns had rudimentary fire laws, but only rarely did these laws call for measures aimed at helping building occupants escape from a fire. At this time, when few buildings exceeded three stories, occupants could be rescued from the windows or could even survive a jump. Around mid-century, the sizes of new city buildings began to increase, and the greater height and volume, when combined with larger numbers of people inside, complicated rescues from the outside. Ladders could not reach the fifth or sixth stories. Buildings had to contain means so that people could get away safely in the event of a fire.

If landlords had to be compelled to make their buildings, in which they had a financial stake, more fire-resisting, then they certainly would not voluntarily invest in measures to protect the people inside their buildings. American legal tradition relieved owners of responsibility for deaths and injuries resulting from fires in their properties, unless the owners personally caused the fire. They had no obligation to provide emergency ways out—redundant facilities, like a second stairway that occupants could use in case smoke or flames blocked the main one, or extra-wide corridors, required to enable many people to leave all at once. Such facilities, unnecessary for daily comings and goings, added to the cost of the building and also took up valuable space. Moreover, the public did not demand such facilities. Some office tenants preferred to rent space in fireproof buildings to better protect their personal property, but there is no evidence that they noticed the number and characteristics of the stairways. Many of the early fireproof skyscrapers—ten or more stories tall—contained only one stairway, although several elevator cars. People going to the upper floors of high-rise buildings ordinarily took the elevators and might not even know the location of the stairs. What was sufficient for normal use, however, might not be adequate in an emergency. People needed to have at least two ways out from a building. To the layperson, these redundant facilities seemed wasteful: why have extra stairs in an elevator-building that no one uses? It usually took a deadly fire to convince the public that such facilities were necessary and that building owners should be compelled to provide them.

One of the earliest built-in kinds of alternative exit was the roof hatch. In the late 1700s, a mutual fire insurance company, the Philadelphia Contributionship, required its policyholders to put trap doors in their roofs. Charlestown, Massachusetts, across the harbor from Boston, took up this idea in 1810, when it called for every building to have roof hatch (called a "scuttle") as well as a ladder leading to it and a "safe railing on the roof." The purpose of the hatches in this early period was to enable firefighters to reach roof and chimney fires. But the hatches eventually evolved into emergency exits: in 1852, Brooklyn required all buildings within the fire limits to have "a scuttle or place of egress in the roof." Soon, New York City and Boston followed suit, and eventually many cities required scuttles. A feature of the scuttle rules that distinguished them from nearly all other building laws was that they applied retroactively to existing buildings in addition to future construction.[2]

Americans had little to guide them when they sought to regulate exits. London's building laws gave exits slight attention, mainly aimed at helping firemen get to fires rather than helping people inside burning buildings get out. The Metropolitan Building Act of 1844 required that the floors in halls and common areas, and also the stairs, of new "public" buildings (buildings the public could enter) be "wholly supported, constructed, formed, made and finished Fireproof." It also called for dwellings, in some cases, to have noncombustible stairways. However, the act did not require these corridors and stairways to be isolated from the rest of the building by fire-resistive walls or doors, so while they would support the firemen, they still could fill with smoke and flames and become impassable. These requirements for London remained essentially unchanged until 1894.[3]

Members of New York City's chapter of the American Institute of Architects undoubtedly consulted London's laws when they recommended provisions for a new building law for the city, which the state legislature enacted in 1860. But New York's law contained egress rules unknown in Britain, spurred by a growing concern about the safety of and sanitary conditions in the city's mushrooming number of tenement houses.[4] Following a field investigation of the city's tenements, a state legislative committee concluded that most were firetraps. In its 1857 report, the committee urged public regulation of the halls and stairways in tenements in order "to insure easy egress in case of fire."[5]

Although tenement house fires were common events, they usually were quickly extinguished. Nevertheless, the potential for disaster loomed and occasionally the fires did prove deadly. One particularly horrible fire, in the winter of 1860, started in a large tenement house on Elm Street, and, according to the initial press reports, claimed thirty lives. This fire outraged the editors of the *New York Times,* who reproached New Yorkers for fretting over the welfare of slaves in the South while ignoring the safety of their own poor neighbors.[6] Within a couple months, the state enacted a new, comprehensive building code for New York City. An early draft of this law called for making landlords liable for injuries sustained at fires in their buildings, but lawmakers deleted this provision from the final version.[7]

Nonetheless, this 1860 law introduced several novel requirements pertaining to egress, the most important of which was a device that became a signature feature of America's downtowns: the outside fire escape. It required every large apartment house (nine or more units) to have either a fireproof stair-tower or else "fireproof balconies on each story on the outside of the building, connected by fireproof stairs." Lawmakers included such a long description of what is known today simply as a "fire escape" because in 1860, the term *fire escape* did not mean an exterior stairway. Most commonly, it meant a ladder on wheels—the ancestor of a fire department's ladder truck. Tenement landlords, given the choice between putting up simple balconies, or else erecting a separate stair-tower, undoubtedly chose the balconies. Nevertheless, the law shows that its drafters realized the importance of enclosing stairways behind masonry walls in order to keep a fire out.[8]

Because of ambiguity over whether the stairway rule applied to existing tenements or only new construction, lawmakers put language in a new building law enacted two years later stating that the rules applied to all large tenements, those "already erected, or that may be hereafter built." At the same time, lawmakers dropped the stair-tower requirement, and instead of fireproof balconies and stairs, simply required every large tenement to have "a practical fire-proof fire escape."[9] This was the first time the term *fire escape* appeared in a building law. New York City's newly appointed superintendent of buildings, James Macgregor, supported the new egress regulations. With construction starts down because of the Civil War, he ordered his inspectors to concentrate on enforcing the fire escape rule.[10]

When state lawmakers deleted the section of the former law that called for balconies with stairs, and instead called for "fire escapes," they signaled that something besides outside stairways would be considered as fire escapes, although they did not indicate what would be acceptable. Inventors turned out an amazing variety of devices they called fire escapes. The city's building inspectors had little to go on in evaluating their "practical-ity," but still approved such things as baskets on ropes attached to pulleys and chain drop-ladders. With portable devices counting as fire escapes and many owners simply ignoring the law, the fire escape rules had no visible impact on New York City through the 1860s and into the 1870s.[11]

In 1867, in another revision of its building law, New York lawmakers revived the idea of protected egress and required *new* tenements to have fireproof stairs and corridors. Rather than being in stair-towers, however, the stairs were to be inside the building, enclosed in brick walls. These walls were to run straight up through the building, "from foundation to roof," similar to a modern fire stair. In addition, the floors in the corridors were to be fireproof—made of brick arches on iron beams—and the stair treads of iron or stone.[12] In this building code, lawmakers for the first time addressed the issue of emergency egress from buildings other than tene-ments. Buildings of a "public character," including hotels, churches, theaters, schools, public halls, and places of amusement, had to have "req-uisite and proper" facilities to enable people to escape in case of fire. How-ever, the law did not prescribe what constituted requisite and proper; this was left to the judgment of the superintendent of buildings.

In calling for fireproof construction for tenement hallways, New York's legislators imposed a costly requirement, so it is not surprising that they soon retreated. In an 1871 revision of the code, they relaxed the fireproof stair rule: brick corridor walls no longer had to rise to the full height of the building but only to the top of the second-story beams, and, with one ex-ception, hallway floors and stairs no longer had to be fireproof.[13] One pro-vision lawmakers retained and expanded in the 1871 law was the require-ment for fire escapes. Already in 1867, they called for fire escapes on large factories as well as tenement houses; now, every hotel, boarding house, of-fice building, and factory in which people worked above the first floor also had to have a fire escape.

In dropping the requirement for fireproof stairs and making fire es-capes the all-purpose solution for emergency egress, lawmakers most likely accommodated the preferences of landlords for a cheap solution. Although two means of egress, even if one is an outside fire escape, is better than a single fireproof stairway, two fireproof stairways would be better still, especially in tall buildings or ones with many occupants. Lawmakers might have made different rules for existing buildings and new construc-tion and demanded, for example, two protected stairways in the latter; but they did not, nor did anyone propose it. The public accepted fire escapes, landlords did not oppose them, and thus, fire escapes became the nearly universal secondary means of egress in the United States in the late nineteenth century.

Boston's first modern building law, which was modeled on New York's code, was enacted in 1871. It required "sufficient" means of "ingress and egress" (borrowing words from London's building laws, which inexplicably specified both) in new theaters and similar sorts of buildings. In addition, every tenement and lodging-house, whether new or existing, regardless of size, had to have "a proper fire-escape, or means of escape in case of fire." The law said nothing about exits in workplaces.[14] The next year, after central Boston burned down, state lawmakers enacted an amended building law with egress rules stricter than those in New York's law. Boston's 1872 law required protected egress in a range of buildings: every new railroad station, auditorium, school, hotel, lodging or tenement house, and factory, in which more than twenty-five people lived, worked, or gathered above the first floor, had to have at least one enclosed stairway made of noncombustible materials. In addition, the law mandated fire escapes on any building where "operatives" worked at the third floor or above, a rule that pertained to dry-goods stores with seamstresses working in the upper floors, for example. Finally, this law introduced a measure that became a fixture of American egress rules: it required outward-opening doors for exits in public buildings.[15]

From this point to the end of the century, Boston led all other cities in the United States in the thoroughness of its egress regulations. Massachusetts modeled its own 1877 statewide egress law on Boston's rules. This law, the first of its kind in the nation and one that preceded by decades a general building law for the state, had its origins in a familiar source—a terrible fire. In this case, the event was the burning of Granite Mill No. 1, a large textile factory in Fall River, on September 19, 1874. A large stone building (five stories with an attic), the Granite Mill contained standpipes and hoses on each floor, supplied with water from a roof cistern, something the factory mutual companies that insured it required. On a workday morning, a fire broke out in a spinning machine on the fourth floor. A worker grabbed a hose, but no water came out, and while he struggled with it in vain, the fire gained headway. In the excitement, no one sent an alarm to the floors above, where perhaps ninety operatives were at work. Tragically, these operatives, mainly women and children, only discovered the building was on fire when smoke came up the stair-tower—the mill's sole stairway. The ladders on the outside of the stair-tower did not reach the attic level. Once the stairs became impassible, the only ways out for these workers were through skylights that opened onto a steep roof, or by jumping out the windows at the gable ends. The city's firefighters arrived quickly, but they could not help the trapped workers: their ladders only reached halfway up the mill walls. Some workers jumped out of the windows and were killed from the fall; others died inside the building. Twenty-three people perished and scores more were injured. Remarkably, this was the first time that a fire in a mill insured by the New England factory mutual fire insurance companies had resulted in a loss of life.[16]

Unswayed by the public outrage and recriminations that followed the fire, the coroner's jury ruled that the mill owners could not be held responsible for the deaths. The owners did not start the fire and did nothing that contributed to the loss of life and limb; for example, there was no evidence that the stair-tower door had been locked, although one worker claimed otherwise.[17] Massachusetts's Supreme Court affirmed this judgment several years later, in a lawsuit brought by an injured worker against the mill owners. In its 1878 ruling, the court found no negligence on the part of the owners and thus no liability for the complainant's injuries, a finding consistent with the prevailing theory of free contract. As the court wrote, "It is no part of the contract of employment between master and servant so to construct the building or place where the servants work, that all can escape in case of fire with safety." Nor did the employer have an obligation to provide a functioning fire-fighting system; the duty to maintain such equipment lay with the workers.[18]

This second decision, coming several years after the Granite Mill tragedy, rekindled the anger many people felt following the original coroner's jury ruling. Among those who believed justice had not been done was Carroll D. Wright, the recently appointed chief of the Bay State's new and floundering Bureau of the Statistics of Labor. In his annual report, Wright included a draft for a "Factory Act" for Massachusetts, intended to prevent a repetition of the Granite Mill disaster. "No love of gain," he wrote, "should be allowed to put human life at risk. The number of manufacturers who knowingly endanger the lives of their operatives is probably very small in this state; but there are undoubtedly some, and these should be restrained by law."[19] Even the employers who wanted to do right by their workers, Wright argued, needed a law to guide them toward effective means. Although the law was brief and hardly radical in its requirements, nevertheless the state did not enact a version of it until 1877.

While Wright called his bill a "factory act," a more accurate name for the 1877 law, as passed, would be the "emergency egress act," since it dealt mainly with egress measures and covered buildings besides factories. It called for fire escapes on large factories. In addition, all churches, schools, hotels, halls, theaters, and other buildings used for public assemblies had to have "such means of egress as the inspector . . . shall approve." All main doors in regulated buildings had to swing outward. Very importantly, the law ended the master-servant contract theory as it related to escaping from fire: henceforth "any person or corporation violating any of the provisions of this act . . . shall also be liable for all damages suffered by any employee by reason of such violation."[20]

The act gave state inspectors wide latitude to order better egress facilities in any building. In granting this discretion, lawmakers recognized that diversity among buildings—their sizes, occupancies, exposures, and so on—made it difficult to prescribe facilities in advance. But the lack of specifics in the law allowed owners to challenge the inspectors' orders. Ladders and platforms had been features of many New England factories for decades, although—like the first roof scuttles—owners put them up to aid firefighters, not for emergency escape. Nevertheless, owners felt they should qualify as fire escapes for the purposes of the law, and Massachusetts's 1877 factory act specifically enjoined inspectors from ordering owners to replace them, "unless such change is necessary for the protection of human life." This led to clashes between inspectors and owners over the adequacy of the ladders.

To put an end to such arguments, in 1879, the state's chief inspector prepared specifications for a proper fire escape. Designed to show what his department considered acceptable, the specifications defined an "ordinary fire-escape" as an outside iron stairway consisting of railed balconies and stairs. The balconies were to be placed by the windows, almost level with the windowsills to allow occupants to step out on them rather than having to jump down. The specifications also detailed acceptable dimensions for the balcony and stairs along with methods for attaching the structure to the wall. The specifications would produce a sturdy, outside iron stairway. Massachusetts lawmakers accepted this standard: in an 1880 amendment to the law, they eliminated the language that allowed owners to use old ladders and expanded the scope of the law so that it applied to a greater number of buildings. All buildings covered by the law had to have inside or outside "stairways"[21] (Fig. 6-1).

In 1879, Pennsylvania passed a statewide fire escape law, following the example of Philadelphia, which in 1876 enacted an ordinance to establish

Fig. 6-1. A substantial balcony-type fire escape, circa 1880, recommended by the Massachusetts District Police to comply with the state's exits law. Unlike most fire escapes, this one is like a normal stairway: the stairs have treads, not rungs, and the incline is near to that of conventional stairs. The final flight is fixed and reaches the ground. Even in Massachusetts, few fire escapes were built to this high standard. "General Specification for Fire-escapes...," copy in *Tenth Annual Report of the District Police,* Massachusetts Public Document No. 32, 1888.

a "Board to Regulate Fire Escapes." The Philadelphia board could order any owner, anywhere in the city, to install a fire escape.[22] Pennsylvania's statewide law covered a long list of buildings (although, unaccountably, not theaters) and required them to have "permanent safe external means of escape therefrom in case of fire."[23] Unlike Massachusetts, Pennsylvania did not establish a special inspection force to implement the law; rather, it gave this responsibility to local boards of fire commissioners and fire marshals. If an area did not have such officials, then a fire escape board was to be composed of school directors. The enforcement side of the law guaranteed failure: not only were the boards unqualified and understaffed, but the law contained no penalties for noncompliance, although it made owners liable for death or injuries from fire in buildings that lacked fire escapes. In 1885, legislators imposed fines, but enforcement continued to be weak except in the case of factories, which the state placed under the authority of state factory inspectors when it enacted a factory law.

The idea of fire escape laws spread nationwide. By the early 1890s, most cities had fire escape rules on the books. In addition, by 1894, sixteen states, including Massachusetts, had factory safety legislation, and most of these laws required fire escapes on industrial buildings.[24] Some states, like Pennsylvania, had statewide fire escape laws. Yet the details of the rules varied from place to place, even for similar kinds of buildings. With one exception, no city required any sort of building to contain two interior stairways. The exception was large factories where combustible goods were manufactured; Chicago, Louisville, Omaha, and Providence required such buildings to have two fireproof stairways.[25]

A serious defect of these rules was that they rarely attempted to scale the exit facilities to the number of people that might be expected to use them. Only in the case of factories did any city's law specify the number of fire escapes or exits for a given number of occupants. For example, two Ohio cities, Cincinnati and Cleveland, required either a fire escape or fireproof stairway for every twenty-five people working on the third floor or above. Chicago and Louisville had similar rules, except they allowed ladders instead of fire escapes or stairways.

Boston's 1885 building law was the first to attempt to scale egress facilities to occupancy in public buildings. In auditoriums, stairways and corridors had to be a minimum of five feet wide, and twenty inches wider for every hundred people the building could hold. Theater owners could calculate widths reliably because one seat equaled one person and temporary seats or standing in the aisles were forbidden. Massachusetts and Boston also pioneered a practical mechanism for limiting occupancy in other kinds of buildings according to exit capacity. An 1888 law for the state authorized inspectors to issue certificates stating the number of people that a building's exit facilities could accommodate. The certificate could be voided if the occupancy increased, or if the owner altered the building materially or reduced the exit facilities, and it had to be renewed every five years. Owners who failed to comply could be denied a license to use the building or, if no license was required, simply barred from using it. In another novel feature, the law required owners to post the occupancy certificate, thereby allowing tenants to police compliance. Under Boston's new building law of 1892, owners or lessees could obtain a certificate "to the effect that his building is provided with safe means of egress," and tenants and employees could request that a building be inspected if the owner did not have one.[26] These seemingly obvious measures were quite advanced for their day, and only Massachusetts and Boston attempted to control occupancy at this time. In other localities, lacking such rules, public officials did not know whether the facilities available in any building would be sufficient for the service they might have to provide.[27]

Although not the first type of structure to be regulated with respect to egress, theaters soon were covered by the most comprehensive rules. With stages and workshops filled with combustible material—paints, painted scrims, ropes, bits of wood—and performances illuminated by candles and gaslights, theaters regularly caught fire. But these fires, like most tenement and factory fires, usually were stamped out before causing much damage. Still, some got out of control: the German civil engineer August Foelsch, who studied theater fires, counted about five hundred fires in Europe and the United States from 1750 to 1877 that left theaters in complete ruins. Several theaters burned multiple times; New York City's Bowery Theatre was rebuilt five times. In honor of its reconstruction after an 1836 fire, the managers of Venice's opera house named it *La Fenice* (The Phoenix). Many fire underwriters considered theaters too risky to insure. The risk that underwriters feared was the destruction of the theater building and its contents, not loss of life, since fire insurance policies did not cover liability for death and injuries to theatergoers.[28]

Although most theater fires broke out either before or after a performance, those that started during a performance caused appalling numbers of casualties. It was a theater fire in 1811, which claimed seventy lives, that helped turn Robert Mills into a prophet of fireproof construction. Theater owners and managers could not be held responsible for casualties in cases

of fire that started accidentally, and most deadly theater fires were "accidental," typically sparked by lights too close to the scenery or other material on the stage area. A few of America's early theater designers tried to
address the problem of fire by using noncombustible construction materials. French architects had already tried this: the *Théâtre Français* (1786–
89) was one of the first fireproof buildings in Paris. In the early nineteenth century, French authorities required iron roofs on Parisian theaters.
Meanwhile, in London and Munich, theater managers experimented with
"fireproof" wood. Yet no designer in the early part of the century paid any
attention to measures that would help people escape from their theaters.[29]

Audiences could not rely on theater owners and managers to look after
their safety. Theater proprietors liked to cram in as many paying customers as possible and had little incentive to protect them by, for example, putting in additional stairways and wide corridors, keeping aisles free, or buying extra street frontage so as to put more doors at the main entrance. Had
proprietors been rewarded for taking such steps—if the public would have
paid more for tickets at safe theaters or refused to attend shows in dangerous ones—they might have acted differently. But people in the mood to attend a popular show thought little about the safety of the building. As
AABN noted after a devastating theater fire in 1876, if the public had to
wait for market forces to improve theater safety, they would be waiting a
long time. Local government alone could protect audiences, it argued, "by
enforcing very stringent rules concerning the arrangement, construction,
and care of theatres."[30]

The disaster that prompted AABN's call for theater regulation, a fire in
Conway's Theatre in Brooklyn, December 1876, was America's deadliest
fire up to that time. The fire started on the stage during a performance and
caused a panic. Most of the nearly three hundred people killed in the fire
had been sitting in the gallery, where they had only one narrow, winding
stairway by which to leave. This stairway became jammed with people trying to flee and collapsed under their weight. Yet the Conway, constructed
in 1871, was a relatively new theater, with exit facilities like those at most
other theaters.

Immediately after this fire, a committee of architects from the New
York City chapter of the American Institute of Architects (AIA) drafted a
bill for the regulation of theater construction and management. The proposed law went to the state legislature in 1877, but lawmakers declined to
enact it as a separate law. Later, with minor modifications, the text became

a section of New York City's new building code of 1885.[31] In the same year, Boston amended its building code and included many new theater rules patterned on those in the New York City AIA bill. The laws in the two cities vary in some details: New York allowed narrower corridors than Boston did and also assumed outside fire escapes would be the second way out from upper seating levels, while Boston's law allowed inspectors to order a second interior stairway. Moreover, in Boston, some of the new egress rules applied to "public buildings" generally, not only theaters. New York's law required theaters to have open passageways along both sides of an auditorium, a rule borrowed from Europe. With a few exceptions, American theaters at this time commonly covered the entire lot, whereas prominent theaters in Europe typically were freestanding. An important similarity in the codes of Boston and New York was the requirement that future theaters be fireproof. Boston's law called for theaters to "be of fire resisting construction throughout, so far as the nature of its uses will permit."[32] While New York's law contained no such general statement, it specified that all constructive parts of a theater be fireproof. This included the exterior walls; the inside walls that enclosed stairs and separated the auditorium from the lobby and stage; all partitions; the proscenium arch; and the floors, including the main floor, the floor over the lobby, and corridors. Thus, in 1885, theaters joined tall buildings as the only kinds required by law to be fireproof.

Many ideas that became standard features of egress regulation were first introduced in theaters. The idea of posting floor plans in a building to show the location of exits derives from the provision in New York's law that required every theater to post a floor plan along with fire safety regulations. Likewise a theater ordinance under consideration in Chicago at the end of the decade ordered theater programs to include plans of the building, showing stairs and exits; in 1892, Boston adopted this idea.[33] Theaters were the first kinds of buildings to have marked exits, an idea introduced in New York's 1885 law.

The Rise of Occupancy-based Egress Standards

Except in theaters, emergency egress received little serious attention. Localities relied almost exclusively on fire escapes as an alternate means of egress in a fire, but they did not insist that fire escapes meet high standards. Owners put up fire escapes with features that made them difficult, if not dangerous, to use: one could find fire escapes with wooden platforms; straight ladders, or steeply inclined ladders with rungs rather than flat treads; and very narrow passages around stair openings.

The poor quality of many fire escapes concerned housing reformers. New York State's Tenement House Commission of 1900 inspected tenements in Brooklyn's Greenpoint section and in several wards on the Lower East Side of Manhattan, to check for compliance with the city's fire escape laws. In Greenpoint, the commission found that a quarter of the tenements had straight ladders attached flat to the walls between windows, without even a platform or balcony to help occupants get to them. More discouraging than this, 43 percent of the tenements in the area had no fire escapes at all. The commission found better compliance in the Lower East Side: only about 3 percent of the tenement houses there lacked fire escapes. But many fire escapes in this area were at the rear of the buildings, where they were more likely to be cluttered with tenants' possessions or inaccessible; others were of substandard design, with straight ladders between balconies rather than inclined stairs.[34]

The commission argued that the city needed an agency that could concentrate on the health and safety problems of tenements.[35] The state legislature established a separate Tenement House Department in 1901, implementing many of the commission's recommendations. Among these were detailed specifications for fire escapes, which put an end to perilous vertical ladders between balconies. Still, they left much to be desired: a fire escape could have stairs at a sixty-degree angle, much steeper than conventional stairs, with a nine-inch rise. Indeed, the inside stairs in tenements could have an eight-inch rise, which also was steeper than what builders' handbooks recommended.[36] People often have difficulty negotiating steep stairs in a hurry.

But even substantial fire escapes built like standard stairways had drawbacks. For one, smoke and flames lapping out of windows in the vicinity of the fire escape could prevent people from descending. Fire escapes in northern climates might be covered with snow and ice in the winter, which made them treacherous. Tenants used fire escape balconies for storage. And fire escapes were sometimes difficult to reach at all: the windows that led to them might be hard to open, blocked off, or not accessible to the public.[37]

One of the few improvements in fire escape design, intended to solve these problems, was the "smokeproof tower." Introduced in Pennsylvania in the late 1880s, it consisted of a completely enclosed stairway that users entered by first going outside on a balcony. The balconies either projected past the building line or were recessed in the building wall, this latter type called a "vestibule tower" (Fig. 6-2). Since the door to the tower opened to the outside, the stairway was unlikely to fill with smoke or flames from a fire inside the building, hence the name "smokeproof." Philadelphia was the first, and probably the only, city to require this type of fire escape: the city's building law of 1899 called for large stores and factories, three or more stories high, to have at least one "tower stairway, completely enclosed, on the interior of the building." The smokeproof tower is essentially a modern fire-protected stairway, although accessed from the outside rather than inside a building; and it would have been a much better emergency exit than the iron fire escapes that were routinely installed on high buildings, theaters, and hotels even though they were unsuited for the service they might have to provide. Yet, except in Philadelphia, the only kinds of buildings in which smokeproof towers were installed were factories.[38]

Besides the question of the best design of egress facilities was the issue of quantity: how many stairways, of what size, were necessary for buildings of various uses, heights, floor areas, and populations? An early attempt to address the quantity question was made in 1891 by a committee brought together by the National Association of Fire Engineers to write a model building code for American cities. Although it failed to prepare a full code, the group agreed that the number of stairways in a building should increase as floor area increased: it recommended that new buildings, three stories or more, have a noncombustible stairway for every 2,500 square feet.[39] New York City adopted this idea for its turn-of-the-century building code. Large stores, factories, and hotels, if over 2,500 square feet, had to have two interior stairways; floors of 5,000 to 10,000 square feet needed three interior stairways; floors 10,000 to 15,000 square feet needed four. The number required for areas larger than this was left for the building chief to determine.[40]

The drafters of New York's tenement laws also took up this idea of occupancy-based egress and required more stairways as the number of rooms, or (in 1903) number of apartments, increased. These laws also distinguished between fireproof and nonfireproof tenements (all tenements seven or more stories had to be fireproof). A nonfireproof tenement house in 1903 had to have two stairways when it contained over twenty-six apartments above the entrance story. Owners could satisfy this requirement with a single, extra-wide stairway, as long as the building did not have

Smokeproof Tower with Outside Balcony Entrance

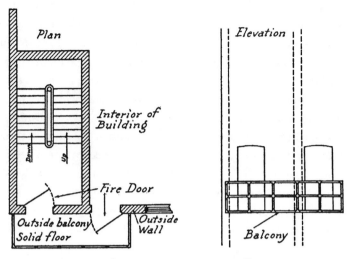

Smokeproof Tower with Vestibule Entrance.

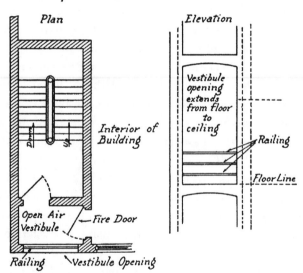

Fig. 6-2. Philadelphia "smokeproof tower" type of fire escape, in plan and elevation. Top shows one with a projecting balcony. The bottom shows one with a recessed (vestibule) entrance. National Board of Fire Underwriters, *Building Code Recommended by the National Board of Fire Underwriters*, 4th ed. (New York, 1915).

more than thirty-six apartments. The law was more lenient for fireproof buildings: it allowed them to have thirty-six units with only one stairway. Two stairways were required for thirty-seven to eighty-four apartments above the entrance story, and three stairways when a building contained over eighty-four units. Tenement House Department officials, along with everyone else at the time, believed people were much safer in a fireproof building. As one explained, "in fireproof buildings, much greater latitude may safely be given than in non-fireproof buildings, and where there are elevators, still further concessions may be granted, as the stairs in such cases are seldom used."[41]

Like fixed theater seats, rooms or apartments served as reliable proxies for the number of persons. Floor area alone, by contrast, was an unreliable indicator of the population of buildings such as lofts, stores, factories, and office buildings. The method of requiring stairways according to floor area had the merit of being easy for owners to understand and for building inspectors to administer, and certainly, the more stairways in a building, the less distance a person would have to travel to reach one. Still, New York City's building department did not know the actual population of any building and, therefore, could not say whether the number of exits in a building was sufficient. The element it lacked was a means to limit occupancy according to the capacity of exits. Although Massachusetts had introduced a mechanism for doing this in its occupancy certification process, no other place took up the idea. But even in Massachusetts, inspectors relied on their experience to judge the adequacy of exit facilities. No systematic data on the subject existed. How long would it take a certain number of people to leave by a given number and size of exit facilities? How long should it take? Almost no one asked these questions.

Fires that start in and spread through the contents of a building create smoke, heat, toxic gases, and flames: the combustion of a building's contents can kill and injure people even if the structure itself never catches on fire. That the occupants even of noncombustible buildings need sufficient means to escape from fire was demonstrated by two deadly fires in fireproof buildings. The second of these fires, in 1911, finally convinced fire protection experts to devote systematic attention to the question of emergency egress.

Although little known, the 1903 fire in Chicago's Iroquois Theatre was the most deadly building fire in the nation's history. Like New York and Boston, Chicago mandated that its theaters be fireproof. This rule had the unintended consequence of lulling owners of fireproof theaters into believing that their buildings were safe. Moreover, the city did not vigorously enforce its building laws: Henry Ericsson, a builder who became Chicago's building commissioner in 1911, recalled that the building commissioner at the turn of the century worked out of the mayor's office. "As a builder I can testify," he wrote, "that the office was not expected to exert any particular influence or control over building as such. This has been the history of the office through successive administrations."[42] This department, headed by George Williams, allowed the Iroquois Theatre to open in 1903 before it was entirely complete—specifically, before all the fire safety measures required in Chicago's building ordinance had been installed. Still, this magnificent theater had the best possible fireproof pedigree: the George A. Fuller Company, one of the nation's most prominent general contractors, built it, and it contained fireproof floors, roofs, and partitions made by the Roebling Construction Company. The fire broke out during a Christmastime matinee, with perhaps two thousand people, mainly women and children, in attendance. It probably started when a border or drapery on the stage brushed up against a hot spotlight. By law, the stage had to have a fireproof curtain, which could separate it from the auditorium in case of fire. But the Iroquois's asbestos curtain became entangled when the stagehands tried to lower it. Soon fire engulfed the stage, and the actors and stagehands fled through a door at the back; as they left, cold air rushed in, which pushed the smoke and flames out toward the seats. To make matters worse for the audience, a ventilator in the rear of the auditorium pulled the smoke up to the balconies. At first the audience did not realize the extent of the fire, because the partly lowered curtain obscured the stage. But when they did, terrified and in complete darkness, they stampeded for the exits. Although the building had emergency exits, few people tried to use them: as in earlier theater fires, they "moved toward the only exits they knew of—the ones through which they had entered." The fire department arrived about five minutes after receiving an alarm, but too late for hundreds of theatergoers. Five hundred and eighty-one people died at the scene, most suffocated by the smoke and gases, others trampled to death; about twenty more, of the over 250 injured, died later of their wounds. The great majority of the dead had been in the balcony sections: 70 percent of the people in the gallery perished.[43]

Yet the building suffered remarkably little damage. In a report on the fire, the Roebling Construction Company suggested that the high quality of the building's construction contributed to the disaster, because it encouraged the owners and managers to leave out "the fire-fighting apparatus which would probably have been provided had combustible construc-tion been used."[44] Nevertheless, the owners failed to comply with many provisions of the building law: the theater lacked automatic sprinklers, and the skylight over the stage, which should have vented the smoke dur-ing the fire, did not operate. John R. Freeman, the Factory Mutual fire in-surance executive and civil engineer, concluded that if the theater had had automatic sprinklers and a functioning skylight, fewer lives would have been lost. Freeman had been asked to investigate the fire by a Chicago businessman who lost two nieces; he reported his findings in a paper he read at a meeting of the American Society of Mechanical Engineers and published in the Society's *Transactions* and also separately, entitled *On the Safeguarding of Life in Theaters*.[45] Of his many recommendations, he con-sidered venting the stage the most important for saving lives.

Like the Roebling Company and most of the other experts who studied the fire, Freeman focused on construction details and equipment for extin-guishing fires, not on the egress arrangements of the theater. Indeed, he judged the exits and stairways from the gallery to be ample. Yet there was also much evidence that poorly designed egress facilities compounded the loss of life. An investigating committee appointed by the mayor found many deficiencies, including exits without signs; steps in front of doors that caused people to trip and then block the way; and doors fastened with an unfamiliar kind of lock. One broad stairway in the gallery led only to a locked door; about fifty to sixty people died in this trap. The fire escapes proved useless: flames lapped out onto them, and a partly open window shutter blocked one of the fire escape balconies. The English fire protec-tion expert and theater designer, Edwin O. Sachs, conducted his own in-quiry into the fire and judged the overall exit arrangements in the Iro-quois a disgrace. The various seating sections did not have separate staircases; streams of people from different sections converged at passages and landings; and everyone entered and practically all left from doors on only one side of the building. He took the opportunity to emphasize his belief that egress rather than noncombustible materials were most impor-tant in protecting lives: "the question of exit and straightforward planning must be given precedence to those of construction."[46]

In urging the priority of egress measures, Sachs had only theaters in mind, and in fact, Chicago lawmakers quickly enacted a new ordinance for theater and assembly hall construction that included better egress regula-tions. The ordinance called for a separate entrance for each seating level; aisles, doors, and corridors scaled to the number of seats; aisles and corridors that led to exits; and marked exits. It also required theaters to obtain a license, which limited the number of persons that could occupy the theater.[47] These valuable improvements in egress rules applied only to theaters. City lawmakers did not take the opportunity to consider whether the rules might be applied to other sorts of buildings. No one thought about high buildings and the problem of evacuating them quickly in a fire. The fact that tall buildings were fireproof reassured everyone that they were safe.

At the end of the century, a new kind of building appeared in New York City: the skyscraper factory. In Midtown and Lower Manhattan, tall loft buildings sprouted up, and while they may have been intended as warehouses, they attracted manufacturers as tenants, especially garment manufacturers. The loft buildings contained only as many stairways as New York City's building law required, a number based entirely on floor area. Thus, buildings of the same size had the same number of stairways whether they contained wholesale goods and a few stock clerks, or manufacturing shops with hundreds of employees. State factory inspectors had the authority to order owners to install fire escapes on overcrowded factories, if they found exits insufficient; however, they had no jurisdiction in New York City. Rather, the city's building inspectors enforced all rules pertaining to buildings, including the fire escape rules, although their experience was in construction, not workplace safety.

New York City's fire chief in the early twentieth century, Edward Croker, was among the few fire protection experts to warn that high buildings, even fireproof ones, were dangerous for people in a fire. As he explained his concerns to a grand jury investigating a deadly fire in 1911, "There are few fire escapes in any of these office and loft buildings of the so-called modern fireproof type." He believed that such buildings did not deserve to be called fireproof once tenants filled them with combustible things: the buildings "are safe so far as property damage is concerned, but not so far as human life is concerned."[48] Reform groups and labor unions also urged government to take steps to better protect workers from fires. In 1908, the New York Federation of Women's Clubs lobbied the City's Board of Aldermen to enact a law requiring fire drills in factories, but the board declined to do so.[49] Up to this time, only secondary schools had adopted the new idea of having fire drills. Then in 1910, following a deadly factory fire in Newark, the federation brought another bill, which called for weekly fire drills in factories, to the New York state legislature. Again, it failed to pass. Although wages were the main issue in the shirtwaist workers strike of 1909, workers also complained about dangerous working conditions, as did the unionized cloak and suit workers who struck the following year. A group formed to investigate the latter union's grievances discovered problems with exits in the majority of the loft workshops they inspected, the principal ones being too few fire escapes and doors that opened inward, contrary to the state factory law.[50] Yet lawmakers could not be persuaded that the city needed new regulations.

The event that gave birth to the systematic study of how to safeguard human life in building fires occurred on March 25, 1911, in one of the sky-scraper factories that Fire Chief Croker warned about. In the evening of that day, a fire started at the Triangle Shirtwaist Company, a blouse manu-facturer, which occupied the eighth, ninth, and tenth floors of the Asch Building near Washington Square in New York City. There was some de-lay in notifying the fire department and by the time help arrived, the fire had engulfed Triangle's shop. The fire department could not fight the fire from the ground and its ladders could not reach the eighth floor; nor could the firemen get up to the fire in time, since the elevators had stopped func-tioning and fleeing workers jammed the stairs. The fire simply burned it-self out.

At ten stories, the Asch Building was required by law to be fireproof, and indeed, it was a standard iron, steel, and tile type of building. Yet the structure failed to contain the fire, which spread past the eighth floor where it started by lapping out the windows and entering through broken windows to the floor above, and by traveling up one of the stairways. The Asch had a standpipe with hoses at each floor, but this system may have been inoperative, and anyway, the factory workers had not been trained to use it. Nothing impeded the fire, which burned with fearful intensity. De-spite this, the structure held up well. In an account of the fire the follow-ing day, a reporter marveled that the building "shows now hardly any signs of the disaster that overtook it. The walls are as good as ever; so are the floors; nothing is the worse for the fire except the furniture."[51] Except too, he went on to say, for the 141 workers killed in the fire.

The Triangle fire, although far less deadly than the Iroquois fire, was more infamous, partly because of the horrifying way many workers died. About six hundred people were at work on Triangle's three floors on the day of the fire. Probably a cigarette discarded among the fabric scraps on the floor sparked it; the blaze then raced through the blouses hanging on lines above the sewing machines, and along the wood floors and sewing machine cabinets soaked with machine oil. The elevators—the way workers normally came and went—ceased operating soon after the fire started. Apart from the elevators, each floor had two stairways and an iron fire escape. In addition, the top floor, the tenth, had an extra stairway that led to the mandatory roof exit, although apparently few workers on the ninth floor knew about it or thought to escape by it. After the fire, workers charged that the owners had locked one of the stairway doors on the ninth floor. Moreover, the stairway doors opened inward, contrary to the law, and became blocked with bodies. The fire escape failed quickly: it became overcrowded when fire bursting out the windows along the balconies blocked workers from going forward and collapsed. For many workers, there was no way out. Trapped, choked by the smoke and scorching heat, they crowded along the windows on the street side of the building: many were pushed out or jumped. Not one survived the fall. The horrible spectacle of bodies smashing on the sidewalks took place before crowds of New Yorkers who gathered around the building but could do nothing to help.

The building law under which the plans for the Asch had been approved required it to have three stairways. However, the owner applied for an exception, and the building department allowed him to put in a fire escape instead of the third stairway. The two stairways that he did build were winding and narrow—so narrow that the stairway doors would have blocked the landings had they opened outward, which is why they opened in, against the direction of travel. By no means could these exit facilities accommodate the large number of people in the building in an emergency. But the building inspectors considered the elevators exits, since there was no rule at the time to avoid them in a fire. Moreover, the building was fireproof. The Asch's owner claimed that he had intended the building to be used as a warehouse. The building department broke its own rules in granting the exception, since the owner faced no "practical difficulties" in complying with the law—supposedly the only grounds for allowing exceptions. Had the building been used as a warehouse, two stairways may have been sufficient.

Yet New York City's law had no mechanism for restricting how a building could be used, or for limiting the number of people inside. New York State's labor law did require 250 cubic feet of space per worker, but this was a health measure—designed to assure that workers had enough air; it could be met in a high-ceiling loft with five feet square per person. An owner did not have to notify the building department if his building changed from, say, warehousing to manufacturing. At the turn of the century, with the availability of building-wide power and small electric gen-

erating engines, manufacturers increasingly sought space in the loft buildings. The big lofts allowed manufacturers to expand operations and gather
many employees under one roof, and insurance companies charged tenants
in fireproof building lower rates. Warehouses were turned into high-rise
factories.

The Triangle fire, unlike the Iroquois, became a turning point in the
history of fire protection. One factor that obscured the lessons the Iroquois
fire might have taught was that it happened in a theater—a kind of building that had for many years been subject to special rules. It was hard to see
past the hazards of theaters to the issue of the danger that combustible
contents posed to people even in fireproof buildings and therefore the
need for sufficient exits in all kinds of buildings. The Asch building was a
typical high-rise, a type going up in great numbers in cities around the
country and used for a variety of purposes: as office buildings, hotels, and
apartment houses as well as warehouses and factories. Yet there were no
rules for emergency exits in tall buildings as such. The Triangle fire grue-
somely exposed this oversight.

A second factor that increased the impact of the Triangle factor was the
ideological and political environment of New York in 1911, so different
from the situation in Chicago in 1903. In New York City, reform groups,
such as the Federation of Women's Clubs, had taken up the cause of
worker protection, and some lawmakers had come to view the issue as
good politics. Fortuitously, a few of these lawmakers held powerful posi-
tions in state government in 1911. For the first time in eighteen years,
Democrats held a majority of seats in both houses of the legislature, as
well as the governor's office. Alfred E. Smith led the Assembly and Robert
F. Wagner, the Senate, both of whom represented New York City. Immediately after the fire, a group of prominent businessmen, political figures,
and reformers organized the Committee on Safety to press the legislature
to enact better worker protection laws. They received a sympathetic hearing from Smith and Wagner, who agreed to sponsor a bill to create a joint
commission to study factory conditions. A bill creating the Factory Investigating Commission passed with little opposition, and Wagner and Smith
became the commission's chairman and vice chairman, respectively.[52]

Although a terrible fire prompted its creation, the commission did not
limit its inquiry to fire safety matters, but investigated a range of occupa-
tional health and safety issues. The fire safety part of the commission's investigation addressed two separate topics—fire prevention how to avert a
fire like the one at Triangle in the first place—and egress. In its earliest re-

port, the commission made commonsense sorts of recommendations with respect to prevention, for example, that factory waste be stored in fireproof receptacles and disposed of regularly; that smoking be prohibited in factories; and that gas jets be covered. But if a fire did break out, what kind of exit facilities would be needed? The commission had the hardest time with questions of quantity: how much stairway capacity was needed for a given occupancy, and how should a building's use and type of construction be factored in when determining this?

The commission's solution to the occupancy/egress problem was to limit occupancy per floor according to the number of people that could fit either within one flight of an enclosed stairway or else in the area behind a fireproof wall or partition. For stairways, the commission recommended allowing fourteen persons per floor for every eighteen inches of stairway width, and one additional person for every three square feet of a landing. Occupancy could increase if story-heights exceed ten feet (because then stairways were longer). A floor could also have such additional numbers of people "as can be accommodated in the smaller of the two spaces divided by the fire wall on the basis of at least three square feet of unobstructed floor area per person." Likewise, adjoining buildings could be counted as areas of refuge if the two buildings had approved openings or balconies connecting them. Twice as many persons were allowed if every floor of a building was equipped with an approved automatic sprinkler system. The total occupancy in a building could not exceed thirty-six square feet per person in a nonfireproof factory and thirty-two square feet in a fireproof building. Owners had to post the allowable occupancy of each floor. Significantly, outside fire escapes and elevators could not be considered in determining the allowable occupancy.[53] The commission put these recommendations in a bill to amend the New York State's labor law, and intended it to apply to New York City along with the rest of the state. However, the rules would cover only factories. While this approach to controlling occupancy could be applied to other kinds of buildings, the commission limited its recommendations to factories.

After the fire, the men who studied fire safety took up the question of emergency egress. In the United States, the principal organization that addressed technical questions concerning fire prevention and protection was the National Fire Protection Association (NFPA).[54] Formed in 1896, the NFPA grew out of an effort on the part of fire insurance interests, manufacturers, and fire safety engineers to establish uniform guidelines for automatic sprinklers. Its mission and membership soon broadened, and it set up committees to develop consensus standards on a host of fire prevention and protection matters—standards that if followed would reduce the nation's fire loss.[55] Safety to life at fires had received little notice in the early years of the organization. Before the Triangle fire, the NFPA had no committee devoted to the question of emergency exits or the safety of building occupants in fires. At its annual meeting in New York City in 1911, shortly after the Triangle fire, members for the first time publicly recognized that protecting lives in fires was a proper topic of concern for the organization. The report of the Committee on Fireproof Construction considered that "the function of a fireproof building is to protect the contents of the building from loss," and this included not only the inanimate contents, but the people, too. The report went on to observe that fireproof buildings, because of their large floor areas and great heights, posed dangers to occupants. Structures may suffer little damage in a fire, still "lives by the hundreds may be lost in so-called fireproof buildings, primarily because the means for promptly extinguishing small fires may be lacking and also the means of exit are inadequate."[56] The experts had begun to appreciate that fireproof construction, while it helped avert conflagrations, could not in itself safeguard lives.

Yet the NFPA did not immediately add emergency egress to its standards-making activities: two years later, in 1913, it authorized the formation of a Committee on Safety to Life. The committee's charge was to recommend new standards for exit facilities, housekeeping, and other matters that would affect the welfare of building occupants in fires. Rather than spread itself over all these topics, the committee chose to focus on egress alone, and began its work by reviewing egress provisions in state and city laws. Not surprisingly, it found existing rules "exceedingly deficient. . . . A number of states . . . have no real legislation upon the subject, many city ordinances are of the most indefinite character, and in some the matter is simply left to the discretion of the fire department and other officials." Moreover, the committee found that the NFPA itself overlooked the human cost of fires: its forms for collecting data on fires did not include lines for reporting deaths and injuries.[57]

The committee focused on two specific problems: the design of "outside stairs"—a name it preferred to "fire escape"—and the appropriate amount of exit facilities for buildings of various uses and sizes. To come up with occupancy-based egress standards, the committee considered New York State's approach as well as a different one adopted by Wisconsin in its 1914 factory law. Rather than the filled-stairway concept used in New York,

Wisconsin assumed that in smaller buildings, stairway area became available as people flowed through and out the door, and therefore more people could be accommodated safely on a floor than could fit all at once in the stairways. However, buildings over six stories were required to provide occupants with uncrowded refuge in enclosed stairways and corridors. Wisconsin also made allowances for building construction and equipment: it permitted more people per unit of stairway in a fireproof, sprinkler-equipped building than in an ordinary building without sprinklers. Operating sprinklers calmed people and encouraged them to leave in a more orderly way.

While it debated standards for occupancy-based egress, the committee also worked on a code for the design of outside stairs. This latter code came to fruition first, and the committee presented a draft of it at the NFPA's 1915 annual meeting; the proposed code covered types, locations, characteristics, lighting, and care of outside stairs. Against the objections of many NFPA members, who felt that fire escapes should simply be prohibited as exits, the committee urged adoption of its standards because so many cities had fire escape rules. These cities at least should be guided in revising their laws so that they required sound and useful fire escapes, such as the committee recommended. In 1916, the NFPA accepted the committee's report and published it as "Outside Stairs for Fire Exits."

Over the next several years, the Safety to Life committee issued occupancy-based egress standards for special kinds of buildings, such as factories, office buildings, and department stores. Then in 1920, at the request of other standards-setting groups, it agreed to oversee the preparation of a model exits code for the nation, which would encompass both design and occupancy standards. An expanded Building Exits Code Committee began work in 1921 and by 1923 had completed the first installment of a comprehensive code, the "Building Exits Code."[58] As with the outside stairs standard, getting agreement on this code proved difficult. Nearly a third of the committee members declined to endorse the tentative code, or voted in favor with reservations. When presented to the full NFPA, many members found fault with it. The code was indeed very long and detailed, but with so many factors to balance, a lengthy, detailed document was unavoidable. A few years later, the committee finished this code; but they continued to revise and update it. Eventually revisions to the code were placed on a schedule, and it evolved from an advisory document to a model law with extensive commentary, today's *Life Safety Code*.

The Triangle fire also led the National Board of Fire Underwriters

(NBFU) to include more information on means of egress in its model codes. Through its Committee on Construction of Buildings, the NBFU had been involved for many years in efforts to get localities to adopt better building codes aimed at averting conflagration. In the 1890s, the NBFU reprinted and circulated a draft building law that had been prepared for New York State, intended for medium-sized cities; it distributed about 25,000 copies of it nationwide.[59] In 1897, the NBFU authorized the committee to prepare new model codes for cities of various sizes. This work went slowly, partly because the committee's chairman believed that New York City's building law, which he helped draft, could not be improved and should serve as a model. By 1904, with a new leader, and spurred by the Baltimore conflagration, the committee took up the task of preparing model laws. Its 1905 *Building Code Recommended by the National Board of Fire Underwriters* was the nation's first model building code. The committee distributed 6,700 copies of it.[60] With the art of building evolving rapidly, the committee continued to revise its code, issuing a second edition in 1907 and a third in 1909. The 1915 edition, which followed the Triangle fire, contained an expanded section on means of egress.[61]

Thus, by the 1920s, cities and states had much more information to help them improve their egress rules. They could find guidelines for exits in the NBFU code; in the labor laws of New York State and Wisconsin; in the NFPA's 1916 recommendations for outside stairs and, after 1923, in its periodically revised "Building Exits Code." What rules cities adopted depended on the relative strength of pressure groups, inclinations of lawmakers, and the physical conditions of a place. Municipal and state rules concerned with egress and life safety changed slowly and unevenly around the country, but as the housing reformer Robert De Forest observed, this was to be expected. In his history of tenement house legislation in New York City, De Forest wrote that although opponents of the 1901 tenement house law succeeded in weakening it through amendments adopted in 1902 and in 1903, he did not feel discouraged. Standards were trending higher; they had to find the proper level along the way: "A standard of housing condition which can be maintained to-day in the great City of New York might not have been maintainable in the same city at an earlier period."[62] Moreover, a standard necessary and acceptable in New York City in 1903 might still be too high for another American city. The interests of landlords and builders, on the one hand, and tenants and the community at large, on the other, had to be balanced in each time and place.

At the same time, the NFPA worked on guidelines for fireproof construction. Chaired by the civil engineer Ira H. Woolson, who also advised the NBFU Committee on Construction of Buildings, the NFPA Committee on Fireproof Construction—after 1913, renamed the Committee on Fire Resistive Construction—sought to define an ideal building, against which actual construction and building codes provisions could be judged. The NFPA adopted the committee's detailed report, "Specifications for Construction of a Standard Building," in 1913, and NFPA members brought these recommendations back to their communities. The report showed that the notion of what constituted a fireproof building continued to evolve. No longer merely a conflagration-stopper, a fireproof building also became a lifesaver. The "Specifications" defined a "Standard Building"—which had the "greatest fire-resistance"—as one "in which the lives of the occupants are properly safeguarded against fire and panic; so designed and equipped that the damage resulting from exposure to fire from within or without shall be reduced to a minimum; and capable of sustaining complete burn-out of its contents without serious injury to its structural members."[63] The safety of occupants now headed the list.

THE INVISIBLE INFRASTRUCTURE OF SAFETY

Suffering through fire is, to a large extent, one of the preventable ills of humanity.

FIRE PREVENTION STUDY,
Alumnae of Bryn Mawr College, 1915

Most building fires are preventable. Those that cannot be prevented can be controlled. Proof of the first point is that the number of building fires in the United States has been falling: in the twenty-year period ending in 1997, the number dropped by about half, from a little over a million to 552,000, even while the nation's stock of buildings increased.[1] Proof of the second point is that even though some of these fires undoubtedly occurred in city centers, none sparked a general fire. The horrors of conflagration, such as those that struck cities in the nineteenth and early twentieth centuries, are unknown today. One of the reasons for this is that downtown buildings today, unlike buildings of the past, are constructed and equipped so that they can contain and resist fires. Over the course of the twentieth century, as new, more fire-resistive buildings have replaced earlier ones, and older buildings have been retrofitted to increase their fire resistance, the chances of conflagration have declined.

This improvement is the result of a long-term process set in motion in the nineteenth century. It involved the development of fire-resistive construction technology, on the one hand, and the enactment of laws that compelled builders to use the new technology, on the other. Both the technology and the laws are invisible to the public. Once a building is finished, the structural fire protection features (save for sprinkler heads) have been hidden. The public knows little about the details of building codes that mandate these fire protection features. The irony is that the achievements of these invisible measures are also invisible. The technology and laws are aimed at *preventing* something from happening. People rarely notice what is *not* there, especially when the absent phenomenon—a big fire—was never a personal threat.

The fire-resistive materials that helped make cities safer also transformed the way buildings are constructed, most importantly, in that they led to the skeleton frame construction. In the late nineteenth century, designers working on fireproof buildings began to use fireproof construction materials in a new way, placing the weight of the building on a frame rather than heavy, bearing walls. Not only did this new approach allow thinner walls, and therefore more natural light, but it also shortened the time needed to construct a building, since work could proceed simultaneously on all floors once the frame was erected, rather than one floor after another, as in bearing-wall buildings. The skeleton frame inaugurated the era of speed in the building industry as much as it did great height. Yet the skeleton frame represented a continuation—an amplification, as one contemporary put it—of the development of fireproof construction. Thus, the desire to build tall did not call the means for doing so into being: all the ingredients for creating the skeleton frame skyscraper already existed.

This is not to suggest the skeleton frame would never have been invented without iron and tile fireproof construction preceding it. But it cannot be doubted that the technological context—the fact that the materials and experience of using them existed—made it easier for designers to go the next step and use these materials in a new way. Moreover, as the fire protection authority Joseph Freitag observed, while a skeleton frame skyscraper does not have to be fireproof (e.g., the floors could be framed in wood), it would be suicidal to build it otherwise. In other words, designers could reasonably imagine ways to build taller buildings because they knew their creations would be fire-resistive. Once the idea of the skeleton frame existed, it came to be used widely because it had other important advantages besides allowing designers to create skyscrapers.

Government played a major role at various points in the development of fireproof construction technology. In the early days, owners largely ignored the system because it cost more than the alternatives. Given this fact, and that owners had other means to protect and indemnify themselves (i.e., fire insurance and municipal fire departments), the wonder is that fireproof construction developed at all. Government helped the technology along, first as a client for fireproof construction at a time when few private owners adopted the system, and second, as a regulator of building construction. Government at all levels built fireproof buildings, but the federal government stands out, being the largest customer for fireproof buildings in the first half of the nineteenth century. In a sense it helped develop the system through its procurement, the sort of effect military spending has often had, when it calls into being things that the private sector would not create but adopts readily once they exist. Thus, the Army's need in the nineteenth century for firearms that could be repaired in the field led to the development of machines for creating uniform, and therefore interchangeable, parts, which laid the groundwork for mass production.[2] Likewise, the Defense Department's need for reliable communication in a national emergency led it to create means for transferring data over networked computers, which eventually evolved into the Internet. In both cases, the specialized requirements of the government prompted it to invest in something that, once started, was taken up by the private sector. Likewise the federal government's requirement for security from fire caused it to adopt, and thereby encourage the development of, a product—rolled beams—for which, in the early days, there was little private demand.

With respect to regulation, by requiring that certain kinds of buildings be fireproof, state and local government helped stimulate demand for non-traditional materials among private owners, which in turn encouraged manufacturers and inventors to enter the business and to introduce improved products. Without these laws, there would have been less fireproof construction and fewer incentives for invention and competition. That public regulation fostered invention may seem surprising, since it is more usual to assume it has the opposite effect: stifling innovation and leading to unimaginative and conservative design practice.[3] But because of the financial disincentives to build fireproof, it is unlikely the process would have advanced as rapidly as it did without the government mandates. Rather than being an economic drag, the pursuit of safety, in this case, proved to be a stimulus to invention.

Municipal building codes not only created demand for the materials of fireproof construction, but the good ones taught designers and builders how to put up safe structures, providing them with an "authoritative and useful hand-book for general reference and a valuable help in many details of special construction."[4] Not all codes did this, of course. Many of the late-nineteenth- and early-twentieth-century codes were disorderly compilations of past and new rules, some obscure, some contradictory, some that increased the cost of construction without achieving the desired objective. Since codes are political as well as technical documents, they inevitably embody political concessions as well. Nevertheless, the building codes in large American cities at the close of the nineteenth century were the most comprehensive and technically sophisticated in the world. The detail in Boston's and New York City's building codes surprised a visiting British architect, who noted that in "the compulsory fireproofing of certain buildings, the provisions of means of escape, and the detailed regulations for safe construction, the American requirements are of much more far-reaching character than our own."[5] With the advent of model building codes, which cities and states gradually adopted, the laws became more uniform and scientifically based.

The codes have not only helped stem great fires, but have made buildings safer for occupants. Like building fires, the number of fire-related deaths and injuries has also trended downward, and in 1997, the number of fire fatalities fell to the lowest level in twenty years—and this while the nation's population has grown. One striking point in the fire fatality figures is the relatively small number of deaths and injuries in nonresidential structures compared with less strictly regulated private homes. In the 1990s, about 80 percent of all deaths and three-fourths of all nonfatal injuries caused by fire each year occurred in homes and apartments; many of the rest occurred in vehicles.[6] Thus, to avoid death or injury from fire, people are better off in a public, code-complying building in a city center, than in their own homes.

Notes

Abbreviations

AABN	American Architect and Building News
AFM	Associated Factory Mutual companies
AIA	American Institute of Architects
AOC	Office of the Architect of the Capitol collections, U.S. Capitol, Washington, D.C.
ARABJ	*Architectural Review and American Builders' Journal*
ASCE	American Society of Civil Engineers
ASME	American Society of Mechanical Engineers
BMMFIC	Boston Manufacturers Mutual Fire Insurance Company, Boston, Mass.
EA papers	Edward Atkinson papers, Massachusetts Historical Society
GPO	Government Printing Office
GSBA	Harvard Graduate School of Business Administration
HABS	Historic American Buildings Survey
HMSO	Her Majesty's Stationery Office
IA	*Inland Architect*
IEES	Insurance Engineering Experiment Station, Massachusetts Institute of Technology
JRF	John Ripley Freeman papers, Massachusetts Institute of Technology archives
JFI	*Journal of the Franklin Institute*
LC	Library of Congress
MHS	Massachusetts Historical Society

MMFIC   Manufacturers Mutual Fire Insurance Company, Providence, R.I.

NBFU   National Board of Fire Underwriters

NFPA   National Fire Protection Association

RG   Record Group

RG 121 Record Group 121, Records of the Public Building Service, National Archives, Washington, D.C.

RIBA   Royal Institute of British Architects

RIHS   Rhode Island Historical Society

TUW   Thomas U. Walter collection, Philadelphia Athenaeum

The Problem of Urban Conflagration

1. Joseph K. Freitag, *Fire Prevention and Fire Protection, As Applied to Building Construction* (New York: John Wiley & Sons, 1921), 6. The word *conflagration* has no standard definition. Usually it means a fire that involves several unrelated buildings and causes substantial property damage.

2. Joseph K. Freitag, *Fire Prevention and Fire Protection*, 7; National Fire Protection Association, *Conflagrations in America Since 1900* (Boston: NFPA, 1951). Counts of buildings and value of property lost vary among sources.

3. John Hall Jr. and Arthur Cote, "America's Fire Problem and Fire Protection," in Arthur Cote, ed., *Fire Protection Handbook* 18th ed. (Quincy, Mass.: NFPA, 1997).

4. The factors that contributed to the decline in urban conflagrations include improved municipal water supply; improved fire alarm systems and alarms in more buildings; better fire-fighting equipment; creation of salaried fire-fighting organizations under government control; widening of old streets; and development of new residential and commercial districts at lower densities. While all these played a role, many fire protection experts in the past considered improved buildings to be the most important factor.

One modern study has looked into the relative importance of these factors. It concluded that improved construction, along with larger house lots (lower density development), were the most significant (L. E. Frost and E. L. Jones, "The Fire Gap and the Greater Durability of Nineteenth Century Cities," *Planning Perspectives* 4 [September 1989]: 333–47). However, this conclusion is based on reasoning—the presumption that since fire fighting is ineffective in a conflagration, it could not have been as consequential as other factors. Another study that touched on this question found that in the Pacific Northwest, urban fires began to decline only after cities there adopted stricter building codes (Daniel Turbeville,

Cities of Kindling: Geographical Implications of the Urban Fire Hazard on the Pacific Northwest Coast Frontier, 1851–1920 [Ph.D. diss., Simon Fraser University, 1985]).

5. NFPA, *Conflagrations in America Since 1900,* 15.

6. Distribution of fires, John Weaver and Peter De Lottinville, "The Conflagration and the City: Disaster and Progress in British North America During the Nineteenth Century," *Histoire sociale-Social History* 13 (November 1980): 417. Using data from six Western European countries, Joseph Freitag calculated that Americans lost ten times more property per capita than Western Europeans (*Fire Prevention and Fire Protection,* 11–12).

7. C. C. Knowles and P. H. Pitt, *History of Building Regulation in London, 1189–1972* (London: Architectural Press, 1972), 18; R. E. H. Read, *British Statutes Relating to Fire 1425–1963* (Borehamwood, Herts.: Department of the Environment, Fire Research Station, 1986).

8. Francis T. Ventre, *Social Control of Technological Innovation; the Regulation of Building Construction* (Ph.D. diss., Massachusetts Institute of Technology, 1973).

9. I do not hyphenate the word *fireproof* in my text, but when I quote sources, I use the original spelling.

10. Fire protection writers first recommended replacing the term *fireproof* at the end of the nineteenth century. Around 1914, the NFPA proposed the term *fire-resistive* instead. The NFPA's Committee on Nomenclature argued that the indiscriminate use of the term *fireproof* had produced "much misunderstanding, and has often engendered a feeling of security entirely unwarranted" (NFPA, *Proceedings of the Twenty-First Annual Meeting* [1917]: 74).

11. Robert S. Moulton, ed., *Crostby-Fiske-Forster Handbook of Fire Protection* (Boston: NFPA, 1935), 339.

12. The term *semifireproof* had a different meaning in the nineteenth century than it did in the twentieth. In the nineteenth century, semifireproof buildings invariably had wood frames. But in the twentieth century, semifireproof construction came to mean noncombustible construction with a lower fire rating on some elements than was allowed in fireproof construction. *Ordinary construction* was the term for buildings with masonry walls and wood interior structures.

Chapter 1
The Solid Masonry Fireproof Building, 1790–1840

Alfred Bartholomew, *Hints Relative to the Construction of Fire-Proof Buildings* ... (London: John Williams, Library of Fine Arts, 1839), p. 6.

1. For Boston, Walter M. Whitehill, *Boston: A Topographical History,* 2nd ed. (Cambridge: Belknap Press of Harvard University Press, 1968), 8, and Carl Sea-

burg, *Boston Observed* (Boston: Beacon, 1971). For New York, Joseph McGoldrick et al., *Building Regulation in New York City* (New York: The Commonwealth Fund, 1944), 27–28.

2. For example, the 1810 fire law of Charlestown, Massachusetts (Massachusetts Acts of 1809, ch. 44).

3. A kind of interlocking iron plate that covered and protected wood, called "fire-plates" by its inventor, David Hartley, enjoyed some popularity in England in the late eighteenth and early nineteenth centuries. Around this time, Lord Mahon, Earl of Stanhope, proposed putting a mixture of lime, sand, and chopped hay between floor joists to make a fireproof barrier, and later a similar compound was used on the roofs of several important buildings in England. Both Hartley's and Lord Mahon's systems were described in Associated Architects, *Resolutions . . . with the Report of a Committee . . . to Consider the Causes of the frequent Fires* (London, 1793).

4. John Fitchen, *Building Construction Before Mechanization* (Cambridge: MIT Press, 1986), 87.

5. Batty Langley, *The London Prices of Bricklayers Materials and Works* (London: Richard Adams and John Wren, 1749), 246.

6. Jonathan Coad, "Two Early Attempts at Fire-Proofing in Royal Dockyards," *Post-Medieval Archaeology* 7 (1973): 88.

7. Felix-Francois d'Espie, *The Manner of Securing All Sorts of Buildings from Fire* (London, 1756?), described in Eileen Harris, *British Architectural Books and Writers 1556–1785* (Cambridge: Cambridge University Press, 1990), 190–91.

8. David Laing, *Plans, Elevations, and Sections, of Buildings . . .* (London: J. Taylor, Architectural Library, 1818), 11.

9. Charles Sylvester, *The Philosophy of Domestic Economy* (London: Longman, Hurst, Rees, Orme, and Brown, 1819), 7–8, footnote.

10. For example, some of the London warehouses built between 1840 and 1860 that were destroyed in the 1861 Tooley Street fire had vaulted cellars (*Building News* 7 [July 5, 1861]: 573). The collapse of King's Warehouse is described in Royal Commission on the Historical Monuments of England, *The London Custom House* (1993), 8.

11. In America, carpenters outnumbered masons by a wide margin (Louise Hall, *Artificer to Architect in America* [Ph.D. diss., Radcliffe College, 1954], 63).

12. Article from the *Philadelphia Monthly Magazine*, 1798, reprinted in Negley Teeters, *The Cradle of the Penitentiary: The Walnut Street Jail at Philadelphia, 1773–1835* (The Pennsylvania Prison Society, 1955), 130.

13. Charles E. Peterson, "Robert Smith, Philadelphia Builder-Architect: From Dalkeith to Princeton," in Richard Sher and Jeffrey Smitten, eds., *Scotland and*

America in the Age of Enlightenment (Edinburgh: University of Edinburgh Press, 1990); Philadelphia Museum of Art, *Philadelphia, Three Centuries of American Art* (1976), 31–32.

14. Negley Teeters, *Cradle of the Penitentiary*, quote from the Act of 1790, 40.

15. Jeffrey Cohen and Charles Brownell, *The Architectural Drawings of Benjamin Henry Latrobe*, vol. 2, part 1 (New Haven: Yale University Press, 1994), 98–112. The prison was eventually completed in 1806 and not according to Latrobe's original plans.

16. Modern architectural writers continue to be puzzled by these combinations of styles. For example, Paul Goldberger described the domed rotunda in New York City's 1830s Doric customhouse as "incongruous" although still excellent (Paul Goldberger, *The City Observed: New York* [New York: Vintage Books, 1979], 21). George Tatum explains the mixture as the architect's attempt to express the values of both Greek and Roman cultures, which make the resulting buildings more a "phase of Romanticism" than classicism (*Penn's Great Town* [Philadelphia: University of Pennsylvania Press, 1961], 54).

17. Talbot Hamlin, *Benjamin Henry Latrobe* (New York: Oxford University Press, 1955), 293–96.

18. From Robert Mills's autobiography, reprinted in Helen M. P. Gallagher, *Robert Mills* (1935; New York: AMS Press, 1966), 159.

19. Talbot Hamlin, *Latrobe*, 419–22.

20. Edward Carter, ed., *The Virginia Journals of Benjamin Henry Latrobe 1795–1798* (New Haven: Yale University Press, 1977), 48.

21. Quote from a letter dated December 31, 1806, in Talbot Hamlin, *Latrobe*, 295. Edward Carter, ed., *The Journals of Benjamin Henry Latrobe 1799–1820*, vol. 3 (New Haven: Yale University Press, 1980), 73, footnote.

22. "New Treasury and Post Office Buildings," U.S. House of Representatives, Report No. 737, March 29, 1838, 28. Letter to the Building Committee of Burlington County Prison, June 4, 1808, *Papers of Robert Mills, 1781–1855*, microfilm (hereafter RM microfilm). In his guidebook, *Guide to the Executive Offices and the Capitol of the United States*, Mills instructed readers that the vaulted Treasury, Patent, and post office buildings were "fire-proof."

23. Photographs of the interiors of the fireproof wings, circa 1895, taken just before the buildings were demolished, are in the collection of the Atwater Kent Museum. Thanks to Tom McGimsey, Historical Architect, for information about the buildings and Ken Jacobs, Architect, for a tour of the cellar.

24. Helen Gallagher, *Robert Mills*, 161, quoting Mills's autobiography.

25. Quote from Mills's report in U.S. House of Representatives, Report No. 737, 1838, 22. Iron sash in the Fireproof Building, Helen Gallagher, *Robert Mills*, 52.

Where the iron sash came from at this time—whether made in the United States or imported—is uncertain. Iron sash could be had in England in the 1810s, and Americans imported English iron sash. By the 1830s, two New York manufacturers produced it (T. U. Walter, *Report ... Upon an Examination of Some Public Buildings of Europe ...*, April 1, 1839, TUW).

26. Cost estimates in Robert Mills's letter to the Treasury Department, December 10, 1833, RM microfilm, roll 6. Figures on actual cost and cost of granite fireproof buildings from U.S. Treasury Department, *Report of the Secretary of the Treasury*, 1861, Table 1, 113.

27. Agnes Gilchrist, *William Strickland, Architect and Engineer, 1788–1854* (Philadelphia: University of Pennsylvania Press, 1950).

28. Historic American Buildings Survey drawings measured April 5, 1939, in the collection of the Philadelphia Athenaeum, and personal visit.

29. Agnes Gilchrist, *William Strickland*, 73. Information about the interior can be found in the records of the Philadelphia Historical Commission.

30. The U.S. Mint at Philadelphia had iron balconies and staircases, and the original design for the Charlotte, North Carolina Branch Mint called for iron balconies (Agnes Gilchrist, *William Strickland*, 81, 84, and 90).

31. Caleb Snow, *History of Boston*, 2nd ed. (Boston: Abel Bowen, 1828), 329. Latrobe's 1807 New Orleans customhouse may have served as a model.

32. *Bowen's Picture of Boston*, 3rd ed. (Boston: Otis, Broaders & Company, 1838). Convict labor made the great increase in stone for construction possible. Prisoners at the state prison in Charlestown cut the stone used in the Massachusetts General Hospital (Massachusetts General Hospital, *Memorial and Historical Volume*, 1921). Prisoners at the Walnut Street jail in Philadelphia cut stone; convicts in the New York prison at Sing Sing dressed the stone used in the Connecticut State Capitol (1827–31), New York University, and New York customhouse. Ohio convicts dressed stone for Ohio's state capitol (Roger Newton, *Town & Davis, Architects* [New York: Columbia University Press, 1942], 158 and 163). Convicts at New Hampshire's state prison in Concord worked stone for out-of-state projects, including buildings in Boston (Donna-Belle Garvin, "The Granite Quarries of Rattlesnake Hill," *Journal of the Society for Industrial Archeology* 20 [1994]: 51). In the South, slaves quarried the stone for South Carolina's state house in the 1850s.

33. Caleb Snow, *History of Boston*, 375.

34. Alexander Parris, Papers, MS collection 2, Massachusetts State House Library, folder 4, plan for the customhouse. Willard designed the courthouse, but Parris built it. A memorial of the architect Richard Upjohn states that in 1833, Upjohn became an assistant to Parris, "then architect of the Boston Court House" (AIA, minutes of the board of trustees, 1878).

35. The Morice Yard in Devonport had vaulted magazines as early as the 1740s

(Jonathan Coad, *The Royal Dockyards 1690–1850* [Aldershot: Scolar Press, 1989], 254–55).

36. The magazine probably had been completed by March 1818, when materials for making black powder arrived at the arsenal (Judy Dobbs, *A History of the Watertown Arsenal, Watertown, Mass., 1816–1967* [Army Materials and Mechanics Research Center], 1977, 5, 10, and 17).

37. Edward Zimmer, *The Architectural Career of Alexander Parris* (Ph.D. diss., Boston University, 1984). William Wheildon, *Memoir of Solomon Willard* (The Monument Association, 1865), 36.

38. Christopher Monkhouse, "Parris' Perusal," *Old-Time New England* 58 (Fall 1967). Chandler to Chandler, City of Portsmouth, New Hampshire, personal property mortgages, 1853. (Thanks to Richard Candee for these references.)

39. William Wheildon, *Solomon Willard*, 27.

40. George D. Seymour, "Ithiel Town—Architect," *Art and Progress* 3 (September 1912): 714–16, and George D. Seymour, "Ithiel Town: Architect: Bridge-Builder 1784–1844," *New Haven* (New Haven: privately printed, 1942).

41. "Speech of Mr. [Levi] Lincoln, of Massachusetts, on the Removal of the Treasury Building," delivered in the U.S. House of Representatives, April 17, 1838 (Washington, D.C., 1838), 3.

42. Jane Davies, "A. J. Davis' Projects for a Patent Office Building, 1832–34," *Journal of the Society of Architectural Historians* 24 (October 1965): 229–51. Constance Green, *Washington, A History of the Capital 1800–1950* (1962; Princeton: Princeton University Press, 1976).

43. Summary of the report of the commission that investigated the security of public buildings after the 1877 Patent Office fire, *AABN* 3 (January 5, 1878): 1.

44. Richard M. Upjohn, *AABN* 38 (November 5, 1892): 80. In addition to the architects cited, Alexander Davis, Town's partner, told Upjohn that William Ross "designed the building as built." See also "The Architecture of 26 Wall Street," U.S. Department of the Interior, National Park Service, Manhattan Sites. The building is now Federal Hall National Memorial.

45. "The Will of the Late Stephen Girard," reprinted from Girard Will Case, 1967 (printed by the College), 13.

46. Bruce Laverty, *Girard College Architectural Collections* (The Athenaeum of Philadelphia, 1994).

47. Agnes Gilchrist, *William Strickland*, 86.

48. I. N. P. Stokes, *Iconography of Manhattan Island, 1498–1909*, vol. 5 (New York: Arno Press, 1967), January 18, 1836.

49. Although changed on the inside, the original facade survives with its remarkable monolithic columns and granite-faced clock. Willard, Rogers's friend and

colleague, supplied the granite from a Quincy, Massachusetts quarry (W. Wheildon, *Solomon Willard,* 233).

50. *Sketches and Business Directory of Boston and Its Vicinity for 1860 and 1861* (Boston: Damrell & Moore & Geo. Coolidge, 1860).

51. On the Capitol, John Latrobe, "The Capitol and Washington at the Beginning of the Present Century, An Address . . . Before the American Institute of Architects in Washington, D.C.," November 16, 1881, 14. On the Treasury building, Mills to Secretary of the Treasury [Louis McLane], April 6, 1833 (RM microfilm, roll 6). When asked for his opinion on the repairs required after the Treasury fire, he reported that the "fireproof wing received little or no injury from the fire, but the main building . . . suffered total demolition." On the 1877 Patent Office fire, *New York Times,* September 25 and 26, 1877, both page 1. For Baltimore: "Detailed Studies of Fireproof Buildings in the Baltimore Conflagration," *Engineering News* 51 (February 25, 1904): 169–73; Roebling Construction Company, *The Baltimore Fire; the Iroquois Theatre Fire* (New York, 1904), 8. Mills's "Fireproof Building" in Charleston stood outside the paths of two great fires in that city, in 1838 and 1861; it did survive the earthquake of 1886.

52. Thomas U. Walter, *A Description of the Girard College for Orphans . . .* (Philadelphia, 1848).

53. "Speech of Mr. [Levi] Lincoln . . . on the Removal of the Treasury Building," April 17, 1838, 5. The corridor was open at both ends when he made his comments.

54. Jacob Abbot, *The Harper Establishment* (1855; Hamden, Conn.: Shoestring Press, 1956), 25.

55. Edwin Betolett, "Fireproof Construction in Philadelphia," *Insurance Engineering* 2 (1901): 200.

56. Betsy Woodman, *A Customhouse for Newburyport (1834–1835)* (Privately printed, 1985), 5.

57. U.S. House of Representatives, Report No. 737, 1838, 23.

58. Letter from Raymond L. Beck, Historian and Curator, North Carolina State Capitol, March 9, 1998.

59. *AABN* 38 (October 29, 1892): 69.

60. Mills to Secretary of the Treasury William Duane, August 14, 1833, in RM microfilm.

61. U.S. House of Representatives, Rep. No. 737, 1838, 21–25. The purpose of this investigation was to embarrass President Jackson by suggesting that the man he appointed was unqualified, as much as to uncover any defects in the building.

62. Information on the First Bank obtained from a personal visit and John Platt et

al., *Historic Structure Report, First Bank of the United States, Independence National Historical Park, Pennsylvania,* Historic Preservation Division, U.S. Department of Interior, April 1981. The vaulted basement survives. The fireproof room in the Arch Street Meeting House is mentioned in James O'Gorman et al., *Drawing Toward Building* (Philadelphia: Pennsylvania Academy of the Fine Arts, 1986), 49–50.

63. "An Act providing for the Safekeeping of the records . . . ," Massachusetts laws of 1812, ch. 165.

64. Isaiah Rogers designed the fireproof addition to the Massachusetts statehouse, which has been demolished (Sinclair Hitchings and Catherine Farlow, *The Massachusetts State House: A New Guide,* 1964). The fireproof rooms in St. Paul's Church still exist.

65. George A. Frederick, "Recollections," New York, 1912, 10–11, manuscript in the collection of the Maryland Historical Society.

66. According to two men with long experience in office building management, brick vaults were "one of the earliest services offered by office buildings" (Earle Shultz and Walter Simmon, *Offices in the Sky* [Indianapolis: Bobbs-Merrill, 1959], 29).

67. A fire insurance inspector wrote that many buildings erected in Boston after the 1872 fire had floors protected with layers of mortar (John E. Whitney, *AABN* 19 [May 22, 1886]: 251).

68. Charles Jewitt, "Appendix to the Report of the Board of Regents of the Smithsonian Institution, Containing a Report on the Public Libraries of the United States," January 1, 1850. Jane Davies, "Wadsworth Atheneum's Original Building . . . ," *Wadsworth Atheneum Bulletin* (Spring 1959): 16–17. (Thanks to John Teahan, Wadsworth Atheneum, for information about the building.)

69. *Sketches of Yale College* (New York: Saxton and Miles, 1843), 100–101. (Thanks to Yale Archives for this reference.) George Seymour, *New Haven,* 219–30.

70. Lydia Sigourney, reprinted in George Seymour, *New Haven.*

71. *A Description of the Tremont House* (Boston: Gray and Bowen, 1830).

72. James O'Gorman et al., *Drawing Toward Building,* 76–77.

73. Specifications, West Chester, Pennsylvania Court House, February 25, 1846, Philadelphia Athenaeum, T. U. Walter collection, papers, box 8. T. U. Walter, "Report of T. U. Walter," December 23, 1851, Senate Ex. Doc. No. 33, 32nd Congress, 1st session. Dexter's drawings are in the collection of the Boston Athenaeum.

Chapter 2
The Iron and Brick Fireproof Building,
1840–1860

1. Harry C. Brearley, *Fifty Years of a Civilizing Force* (New York: Frederick A. Stokes, 1916), 238.

2. Quote from the 1802 *The Beauties of England and Wales*, in H. R. Johnson and A. W. Skempton, "William Strutt's Cotton Mills, 1793–1812," *Transactions of the Newcomen Society* 30 (1955–56 and 1956–57): 180. A recent survey of the literature can be found in Keith Falconer, "Fireproof Mills—The Widening Perspectives," *Industrial Archaeology Review* 16 (Autumn 1993): 11–26.

3. The floors in Robert Mills's Pennsylvania State House Row weighed an estimated 213 pounds per square foot. The weight of segmental brick arch floors varied with the size of the bricks and the material used to fill the top of the arch. In England, an iron and brick floor made with English bricks (larger than American bricks), filled on top and covered with a wood finish floor, weighed about 115 pounds per square foot (John R. Freeman, "Comparison of English and American Types of Factory Construction," *Journal of the Association of Engineering Societies* 10, remarks presented September 17, 1890). Around 1860, standard American floor arches made of common brick leveled up with concrete weighed about 70 pounds per square foot, not counting the weight of the beams, finish floor, or drop ceiling (William J. Fryer Jr., *Architectural Iron Work* [New York: John Wiley & Sons, 1876], 91).

4. Peter Wight, "Fireproof Construction and the Practice of American Architects," *AABN* 41 (August 19, 1893): 113.

5. On the Chatham mill, see Jonathan Coad, *The Royal Dockyards 1690–1850* (Aldershot: Scolar Press, 1989), 34, 237–39. The engineer Marc Brunel designed the building with Edward Holl, engineer and architect of the dockyards. (Thanks to B. L. Hurst for sending me a photo of it.) On sugar refineries, see Luke Hebert, *The Engineer's and Mechanic's Encyclopaedia*, vol. 1 (London: Thomas Kelly, 1836), 526–27. As used in textile mills, Colum Giles and Ian Goodall, *Yorkshire Textile Mills, 1770–1930* (London: HMSO, 1992), 64. One extant example is a wing of the Beehive Mill in Manchester, England, constructed in 1824.

6. Thomas Tredgold, *A Practical Essay on the Strength of Cast Iron* (London: J. Taylor, Architectural Library, 1822). For a supposedly more practical version, William Turnbull, *A Treatise on the Strength, Flexure, and Stiffness of Cast Iron Beams and Columns* (London: J. Taylor, Architectural Library, 1832).

7. *Building News* 22 (March 29, 1872): 250. On the Reform Club, see John Olly, "The Reform Club," *The Architects' Journal* 181 (February 27, 1985): 43.

8. Reports on both incidents can be found in the appendixes to William Fairbairn,

On the Application of Cast and Wrought Iron to Building Purposes (New York: John Wiley, 1854).

9. James Braidwood, *Fire Prevention and Fire Extinction* (London: Bell and Daldy, 1866), quote on 48.

10. Quoted in William Fairbairn, *On the Application of Cast and Wrought Iron to Building Purposes*, 130.

11. So says Eaton Hodgkinson in part II of his 1842 edition of Tredgold. See also R. J. M. Sutherland, "Pioneer British Contributions to Structural Iron and Concrete: 1770–1855," in Charles Peterson, ed., *Building Early America* (Radnor, Pa.: Chilton Book Company, 1976).

12. Eaton Hodgkinson, "Theoretical and Experimental Researches on the Strength and Best Forms of Iron Beams," Manchester Literary and Philosophical Society *Memoirs*, 2nd series, 5 (1831): 407–544. Eaton Hodgkinson, *Experimental Researches on the Strength and Other Properties of Cast Iron* (London: John Weale, 1846). An overview of beam design can be found in Ron Fitzgerald, "The Development of the Cast Iron Frame in Textile Mills to 1850," *Industrial Archaeology Review* 10 (Spring 1988): 127–45.

13. Charles Sylvester, *The Philosophy of Domestic Economy* . . . (London: Longman, Hurst, Reese, Orme, and Brown, 1819), 7. Bulfinch's copy is in the collection of MIT's Rotch Library. Baldwin's copy of Tredgold is also at MIT.

14. Zachariah Allen, *The Practical Tourist* (1832; New York: Arno Press, 1972).

15. Eaton Hodgkinson, "Theoretical and Experimental Researches to Ascertain the Strength and Best Forms of Iron Beams," *JFI* 9, new series (1832) and 10, new series (1832), in parts. A. D. Bache, "Report of Thos. Jefferson Cram . . . upon experiments relative to the strength of Cast-iron beams," *JFI* 18 (September 1836): 153–57. By the time Americans began to construct iron and brick fireproof buildings to any extent, in the 1850s, wrought iron I-beams were available and for this reason, Hodgkinson's ideal cast iron girder with an arched profile and parabolic flanges never caught on.

16. Peter Temin, *Iron and Steel in Nineteenth Century America: An Economic Inquiry* (Cambridge: MIT Press, 1964), chapter 3.

17. Thomas Nolan, editor-in-chief, *The Architects' and Builders' Pocket-Book*, 16th ed. (New York: John Wiley & Sons, 1916), 449.

18. *Journal of the American Institute* (1837): 273–74, names four stores in New York with iron columns. William J. Fryer Jr., who managed an architecture ironworks, wrote that cast iron columns and lintels had been "long previously used" before 1848, the date of Bogardus's first complete iron front (*The Architectural Record* 1 [1891–92]: 232). The architect and engineer George M. Dexter substituted iron columns for granite posts in several Boston storefronts at this time

(George Minot Dexter, drawings c. 1830s–40s, in the collection of the Boston Athenaeum).

19. On the history of iron in railroad bridges, see Theodore Cooper, "American Railroad Bridges," ASCE *Transactions* 21 (July 1889): 1–59. Eric Delony, "Surviving Cast- and Wrought-Iron Bridges in America," *Journal of the Society for Industrial Archeology* 19 (1993): 22–23. On the Brownsville bridge, Frances Robb, "Cast Aside: The First Cast-Iron Bridge in the United States," *Journal of the Society for Industrial Archeology* 19 (1993): 58. On cast iron trestles (although the authors do not identify the bridges), James Dilts and Catharine Black, eds., *Baltimore's Cast-Iron Buildings and Architectural Ironwork* (Centerville, Md.: Tidewater, 1991), 7. Robert Fulton recommended constructing iron bridges across canals for the sake of permanence, durability, and (long-run) economy, in *A Treatise on the Improvement of Canal Navigation* (London: I. and J. Taylor at the Architectural Library, 1796), chapter 22.

20. Zerah Colburn, "American Iron Bridges," Institution of Civil Engineers *Minutes of Proceedings* 22 (1862–63): 540.

21. Peter Temin, *Iron and Steel in Nineteenth Century America*, 20–21; Robert Jewett, "Solving the Puzzle of the First American Rail-Beam," *Technology and Culture* 10 (1969): 382.

22. Around 1850, an important improvement in the cupola was introduced, "the drop bottom, without which no efficient cupola could operate today" (Bruce Simpson, *History of the Metal Casting Industry*, 2nd ed. [1948; Des Plaines, Ill.: American Foundrymen's Society, 1969], 192).

23. The classic article on this subject is Turpin Bannister, "Bogardus Revisited, Part I: The Iron Fronts," *Journal of the Society of Architectural Historians* 15 (December 1956): 12–22. On the beginning of the architectural iron industry in various cities, see James Dilts and Catharine Black, eds., *Baltimore's Cast-Iron;* Ralph Chiumenti, *Cast Iron Architecture in Philadelphia* (Old City Civic Association of Philadelphia and Friends of Cast-Iron Architecture, New York, February 1976); William J. Fryer Jr., "A Review of the Development of Structural Iron," in *A History of Real Estate, Building and Architecture in New York City* (1898; New York: Arno Press, 1967). Margot Gayle's introduction to *Badger's Illustrated Catalogue of Cast Iron Architecture* (New York: Dover, 1981) provides an overview of the development of cast iron used in buildings.

24. Henry Ericsson, *Sixty Years a Builder: The Autobiography of Henry Ericsson* (1942; New York: Arno Press, 1972), 184. James Bogardus, the pioneer cast iron front manufacturer, wrote that the San Francisco buildings were *not* cast iron, but rather were sheet iron wrapped on wood or iron frames (James Bogardus with John W. Thomson, *Cast Iron Buildings: Their Construction and Advantages* [New York, 1856], 12).

25. Peter Wight, "Recent Fireproof Building in Chicago," *IA* 5, extra number

(April 1885): 52. Quote listing the advantages of iron fronts from Mr. Ayres, representing Bogardus's firm, in *Scientific American* 2, new series (1860): 295. Structural cast iron elements have survived intact through many devastating fires.

26. Alexander Parris, "Plans of Buildings and Machinery Erected in the Navy-Yard Boston from 1830 to 1840," National Archives, T-1023. History of the building, Edwin Bearss, *Historic Resource Study, Charlestown Navy Yard, 1800–1842,* vol. 2, U.S. Department of the Interior, National Park Service (October 1984), 736–37, 951–63. Although the building still stands, the fireproof floors are gone. Parris apparently proposed to build the first and second floors fireproof as well (Helen Davis et al., "Alexander Parris: The Years with the Boston Naval Shipyard," [February 1974], 16, in the collection of Boston National Historical Park). Thanks to the National Park Service for helping me get access to the building.

27. Constance Greiff, *John Notman, Architect, 1810–1865* (Philadelphia: Athenaeum of Philadelphia, 1979), 16–17 and 69–71. In 1817, Playfair designed a fireproof room for a museum at Edinburgh University, which had floors made of iron plates on iron girders.

28. U.S. House of Representatives, Report No. 737, 1838, 32. On the Boston customhouse, see Alexander Parris, Papers, MS collection 2, Massachusetts State House Library, folder 4.

29. Constance Greiff, *John Notman,* 119–20.

30. "Fire-Proof Building for the War and Navy Departments," 29th Congress, 1st session, H.R. Executive Doc. No. 186 (1846), 23.

31. "On the Construction of Fire-proof Buildings," *JFI* 11, 3rd series (1846): 153–57, from *Glasgow Practical Mechanic and Engineering Magazine.*

32. "Report, Commission on Public Buildings and Grounds," H.R. No. 102, 32nd Congress, 1st session (serial set no. 656). See also National Park Service, Historic American Building Survey, DC-392, at the Library of Congress. Remarkably, the beam soffits, arches, tie-rods, and other details are still visible.

33. Charles C. Jewett, "Appendix to the Report of the Board of Regents of the Smithsonian Institution, Containing a Report on the Public Libraries of the United States," January 1, 1850, 91–92. A contemporary described the roof as made of "truss beams . . . constructed of cast-iron pipes, in a parabolic form, on the same plan with the iron bridges in France and other parts of Europe, with a view to secure lightness and strength" (from a report reprinted in *Literary World,* September 22, 1849). Saeltzer, a German immigrant and perhaps a student of Karl Schinkel, may have been influenced by the fireproof library in Schinkel's Bauakademie in Berlin (1831–36).

34. "Report of the Architect of Public Buildings," December 1, 1852, S. Doc. 1, 33rd Congress, 2nd session (serial set no. 658), 586. "Specifications for the Iron Work for the Repairs of the Interior of the Congressional Library," May 8, 1852,

and contracts with Janes, Beebe & Company, June 21, 1852, and April 5, 1853, T. U. Walter collection, papers, Philadelphia Athenaeum, box 13. "Report of the Architect of the Extension of the Capitol, Message from the President," February 12, 1852, House Ex. Doc. no. 60 (serial set no. 641).

35. Robert Dale Owen, *Hints on Public Architecture* (1849; New York: DaCapo Press, 1978), 100.

36. Robert Dale Owen, *Hints on Public Architecture,* 107.

37. "Report of the Building Committee," *Annual Report of the Board of Regents of the Smithsonian Institution* for the year 1854, 71.

38. "Report of the Building Committee," *Annual Report of the Board of Regents of the Smithsonian Institution* for the year 1852, 87. Information on the history of this building comes from the *Annual Reports,* 1849–54, and Robert Esau, "Fear of Fire: An Investigation into the Fireproofing of the Smithsonian Institution Building c. 1846–1890," Smithsonian Institution, August 10, 1990. In addition to a wood-framed roof, the regents kept the nonfireproof wings that flanked the central block, completed at the time of the accident, and the wooden stairways. The roof caught fire some ten years later.

39. "Specifications for a Custom-House to be built at San Francisco, California," March 22, 1851, National Archives, RG 121, entry 26, San Francisco Custom House, box 1116.

40. Edwin Bearss, *Historic Resource Study, Charlestown Navy Yard,* vol. 2, 768; Henry T. Bailey, "An Architect of the Old School," *New England Magazine,* new series 31 (November 1901–2): 326–49. Bryant to Acting Secretary of Treasury Hodge, March 8, 1852, National Archives, RG 121, entry 26, San Francisco Custom House, box 1116.

41. On Hayward, Bartlett, see Ferdinand Latrobe, *Iron Men and Their Dogs* (Baltimore: Ivan Drechsler, 1941), 24. On the Massachusetts statehouse extension, *The Capitol of Massachusetts Showing the Enlargement Erected in 1853 & 54,* portfolio of architectural plans in the Fine Arts Department of the Boston Public Library. On Springfield city hall, *Exercises at the Dedication of the New City Hall,* January 1, 1856 (Springfield, 1856).

42. *Badger's Illustrated Catalogue,* 10.

43. Isaiah Rogers's Diaries, transcription vol. 6, Avery Library, Columbia University.

44. Drawing dated 1863. Boston Public Library competition entries are in the Proposals and Plans for the Boston Public Library, 1855, Boston Public Library, Print Department.

45. Joseph K. Freitag, "Steel Buildings and Our Steel Industry," *Fireproof Magazine* 3 (October 1903): 35. This particular form of floor framing, the $\top$ girder and $\curlyvee$ joist, was used by David Mocatta in the 1841 West London Synagogue in Eng-

land (Robert Thorne, ed., *The Iron Revolution*, Essays to Accompany an Exhibition at the RIBA Heinz Gallery, June–July 1990, 4, plate). The Architectural Iron Works in New York City manufactured a Y-shaped "vault beam," so-called because the angled web was designed to start a brick vault (*Badger's Illustrated Catalogue*, plate 53).

46. U.S. Department of Commerce, Bureau of the Census, *Historical Statistics of the United States* (Washington, D.C., 1975), series Y 352–357, 1106. Bowman to James Guthrie, Report on status of projects underway, November 24, 1854, National Archives, RG 121, entry 6. Records of the Public Building Service in the National Archives—RG 121—provide a wealth of information about these buildings. Marine hospitals were civil, not military, buildings, erected for the care of merchant seamen; support for their operations came partly from sailors' wages and the rest from congressional appropriations ("Reports of the Secretary of the Treasury on the State of Finances," H.R. Doc. 3, 33rd Congress, 1st session [1853–54], 20).

47. When Young was appointed supervising architect of the Treasury in January 1852, he became the first person to hold this position. Some writers believe Robert Mills had had the job and that Young replaced him, but this was not the case. Young was a salaried employee of the Treasury Department. Mills, in contrast, was a private architect working on contract. Mills's projects were taken away from him: the Treasury building in Washington went to the Treasury's Office of Construction, and the extensions of the Patent Office, post office, and U.S. Capitol were given to Thomas U. Walter.

48. National Archives, RG 121, entry 26, Cincinnati Custom House, box 179.

49. National Archives, RG 121, entry 26, Cincinnati Custom House, "Specifications for the Custom-House . . . at Cincinnati, Ohio," 1852: "All the girders of the first and second story floorings must be tried and proved before using, by the application of 35,000 pounds, and those of the attic flooring by 22,000 pounds to their centre, their ends being supported, which must in no case cause a deflection therein at the centre of more than ⅜ of an inch. The beams are to be proved in the same way." On British practice, see R. J. M. Sutherland, "The Age of Cast Iron 1780–1850: Who Sized the Beams?," 30, and Lawrance Hurst, "The Age of Fireproof Flooring," in Robert Thorne, ed., *The Iron Revolution*, 39.

50. National Archives, RG 121, entry 26, Cincinnati Custom House, box 179.

51. National Archives, RG 121, entry 26, and "Plans of Public Buildings in the Course of Construction Under Direction of the Secretary of the Treasury," 1855–56. "Report by E. B. White, Jan. 1, 1853," National Archives, RG 121, entry 26, Charleston Custom House, box 133.

52. Did Horton & Macy actually make the castings for this customhouse? Fragmentary evidence suggests the firm did. Isaiah Rogers had already used iron in several Cincinnati buildings: the Commercial Bank, Burnet House hotel, and a

store. Horton bid on the ironwork for the Commercial Bank and presumably was doing this kind of work already (Isaiah Rogers's Diaries, transcription vol. 6, Avery Library, Columbia University).

53. "Reports of the Secretary of the Treasury on the State of Finances," H.R. Doc. 3, 33rd Congress, 1st session (1853–54), 19.

54. A. J. Bloor, in Minutes, American Institute of Architects Board meeting, August 16, 1877, manuscript in AIA library, Washington, D.C. On military engineers, see Todd Shallat, "Science and the Grand Design: Origins of the United States Army Corps of Engineers," *Construction History* 10 (1994): 17–27, and Daniel Calhoun, *The American Civil Engineer* (Cambridge: MIT Press, 1960).

55. George Cullum, *Biographical Register of the Officers . . . of the United States Military Academy at West Point, N.Y.,* vol. 1, 3rd ed. (Boston: Houghton, Mifflin, 1891).

56. Young letter, October 26, 1852, National Archives, RG 121, entry 6.

57. The regulations are printed in "Reports of the Secretary of the Treasury on the State of Finances," H.R. Doc. 3, 33rd Congress, 1st session (1853–54), 278–84.

58. R. J. M. Sutherland, "The Introduction of Iron Into Traditional Building," in Hermione Hobhouse and Ann Saunders, eds., *Good and Proper Materials* (London: The Royal Commission on the Historical Monuments of England, 1989), 48.

59. The earliest instances of wrought iron being used structurally for fire protection include the circa 1785 floors built by Ango and Saint-Fart (iron truss and hollow pots) and the circa 1789 wrought iron joists in a building designed by Nicolas Goulet (Frances Steiner, *French Iron Architecture* [Ph.D. diss., Northwestern University, 1977], 43). These were early instances of a wrought iron and tile system of fireproof construction, described in Charles L. G. Eck's *Traité de Construction en Poteries et Fer* (Paris: J. C. Blosse, 1836).

60. Edwin Clark, *The Britannia and Conway Tubular Bridges* (London: Day & Son, 1850), 137.

61. William Fairbairn, *On the Application of Cast and Wrought Iron to Building Purposes,* quote on vi and 79–81.

62. According to W. K. V. Gale, a student of Britain's iron and steel industry, no British mill rolled wrought iron joists before the 1850s. He wrote that they were never rolled in quantity in Britain and typically were small and so of limited use (*Iron and Steel* [Buxton, Derbys: Moorland, 1977]; "The Rolling of Iron," *Transactions of the Newcomen Society* 37 [1964–65]: 35–46). On the use of rolled iron joists in Fox and Barrett's floor, see Lawrance Hurst, "The Age of Fireproof Flooring," 35.

63. Esmond Shaw suggests diversification was Cooper & Hewitt's motive for entering the beam business (*Peter Cooper and the Wrought Iron Beam* [New York: Cooper Union, School of Art and Architecture, 1960]).

64. Examples are the Norfolk, Virginia and Charleston, South Carolina custom-houses. Construction began at Norfolk in 1853; C&H shipped 112 beams to this project in 1855. Work at Charleston was under way by 1852; beams were sent to it in 1857.

65. Cooper & Hewitt to M. C. Meigs, August 12, 1854, AOC, Capitol Extension—Iron Beams—Girders 1855–71, box 9. Trenton Iron Works succeeded in rolling solid beams before the Assay Office was completed; solid beams may have been used in the top floor. An engineers' report dated August 5, 1873, when repairs were being planned, described the floor framing: "The girders are constructed of 7" and 8" channel iron bolted together, and the 8" floor beams are butting against these girders and fastened thereto with angle irons and bolts" (National Archives, RG 121, entry 26, New York Assay Office, box 286). Were the eight-inch floor beams solid; were they found on other floors; were they installed in 1854 or added later? Unfortunately, I cannot find the answers.

66. Bowman to Guthrie, October 7, 1853, National Archives, RG 121, entry 26, New York Assay Office, box 285.

67. Before they could roll a symmetrical I, Cooper & Hewitt produced beams with the rolls they created in 1849 for a special, deep rail ordered by the Camden & Amboy Railroad, hence the rail shape (Robert Jewett, "Solving the Puzzle of the First American Structural Rail-Beam"; Esmond Shaw, *Peter Cooper and the Wrought Iron Beam*, 35).

68. Cooper & Hewitt to M. C. Meigs, August 12, 1854, AOC, Capitol Extension—Iron Beams—Girders 1855–71, box 9.

69. Quote, Cooper & Hewitt letter to M. C. Meigs, August 2, 1856, AOC, Other Public Buildings, box 14; Cooper & Hewitt to M. C. Meigs, October 29, 1855, AOC, Capitol Extension—Iron Beams—Girders, 1855–71, box 9; letter of June 24, 1857, National Archives, RG 121, entry 6.

70. Samuel Reeves to M. C. Meigs, August 12, 1857, AOC, Capitol Extension, box 9. Some sources state that Phoenix also began rolling beams in 1855 (for example, "Phoenix Steel Corporation, Company History," an introduction to the Phoenix Steel Corp. collection at Hagley Library). However, no contemporary evidence available today supports this date. The earliest documents that mention the availability of Phoenix beams date from February 1857.

71. Patent no. 18,738, dated December 1, 1857.

72. J. K. Freitag, "Steel Buildings and Our Steel Industry," 35.

73. *ARABJ* 1 (1869): 104.

74. Letter of November 11, 1854, National Archives, RG 121, entry 6.

75. Young to Bowman, October 9, 1853, National Archives, RG 121, entry 6.

76. On iron roofs in British mills, Keith Falconer, "Fireproof Mills—The Widening Perspectives," 17. For information on Merrick, see Darwin Stapleton, *The*

Transfer of Early Industrial Technologies to America (Philadelphia: American Philosophical Society, 1987), 22–23. William Strickland illustrated Merrick's roof in William Strickland et al., *Public Works of the United States of America* (London: John Weale, 1841). In 1846, John Notman proposed an ornate iron roof frame for the reconstruction of the fire-damaged Pennsylvania Academy of Fine Arts (Constance Greiff, *John Notman*, 111).

77. The *JFI* carried news of corrugated iron in "Walker's Corrugated Iron Roofs and Gates," 12, new series (1833): 43–45, reprinted from *Mechanics Magazine*. Lefferts's 1854 catalogue is reprinted in Diana Waite, comp., *Architectural Elements, The Technological Revolution* (Princeton: Pyne, 1972). For Lefferts's specifications for the Assay Office roof, Lefferts to Bowman, October 18, 1853, National Archives, RG 121, entry 26, New York Assay Office, box 285. A drawing of the Assay Office roof is in W. Steinmetz's report, October 17, 1873, National Archives, RG 121, entry 26, New York Assay Office, box 286. In 1859–60, Hayward, Bartlett & Company made the iron roof for the original partly fireproof Peabody Institute Building (Archives, Peabody Institute, "Board of Trustees Building Committee, Minutes—Committee Meeting," Box II D 11 12, folder 8).

78. Bowman to Marshall Lefferts & Bro., New York, August 3, 1855, National Archives, RG 121, entry 6.

79. "Report of the Engineer in Charge of the Office of Construction, . . . Sept. 30, 1858," Statement no. 10, Report of the Secretary of the Treasury . . . for the year ending June 30, 1858 (Washington, D.C., 1858), 89.

80. Cooper, Hewitt & Company letter to M. C. Meigs concerning the price of beams: "orders are essential at this time to keep our men from starvation and our good name from discredit. If therefore you know of any good reason why we should take a less price, we shall assent to whatever terms may be offered" (October 7, 1857, AOC, Capitol Extension, box 9). Although the Trenton mill continued to make bars and rails for private customers during the recession, probably the government was the only customer for beams.

81. In July 1857, Phoenix wrote to Bowman announcing that it could supply seven- and nine-inch beams "for use in government buildings"; Bowman ordered some to be sent to Trenton for testing (letter of July 13, 1857, National Archives, RG 121, entry 6). The following month, the firm offered its beams to Capt. Meigs, who was supervising the U.S. Capitol extension (Samuel Reeves to M. C. Meigs, August 12, 1857, AOC, Capitol Extension, box 9).

82. Samuel Reeves to M. C. Meigs, August 12, 1857, AOC, Capitol Extension, box 9.

83. Bowman to James Guthrie, March 18, 1854, National Archives, RG 121, entry 6. The Department requested $3,500 for a "complete series of tests" of iron beams, but it is unclear how much money Congress authorized for testing ("Letter of the Secretary of the Treasury to Hon. Mr. Hunter, Chairman, Com-

mittee on Finance," Senate Exec. Doc. no. 54, 33rd Congress, 2nd session, 1854–55 [serial set no. 752]).

84. Results of both tests were reported by Cooper & Hewitt in a letter to Meigs, August 12, 1854, AOC, Capitol Extension, box 9.

85. Some of test results were attached to the 1855 letter, "Letter of the Secretary of the Treasury to Hon. Mr. Hunter," Senate Exec. Doc. no. 54, 33rd Congress, 2nd session, 1854–55 [serial set no. 752]. Cited in Charles Peterson, "Inventing the I-Beam: Richard Turner, Cooper & Hewitt and Others," in H. Ward Jandl, ed., *The Technology of Early American Buildings* (Washington, D.C.: Foundation for Preservation Technology, 1983): 83.

86. Letters dated July 13, 1857 and December 15, 1857, relative to Phoenix tests, and Bowman to Morris Jones & Company, April 5, 1857, National Archives, RG 121, entry 6.

87. "Results of trial of tensile strength of seven specimens of Wrought iron . . . ," December 23, 1857, AOC, Capitol Extension, box 9.

88. Robert G. Hatfield, "Fireproof Floors for Banks, Insurance Companies, Office Buildings and Dwellings," AIA, *Proceedings of the Second Annual Convention* (1868): 34.

89. Bowman letter, October 8, 1856, National Archives, RG 121, entry 6. The collected drawings were entitled "Plans of Public Buildings in the Course of Construction under Direction of the Secretary of the Treasury," 1855–56.

90. For example, Bowman to S. I. Ravenel of Charleston, October 18, 1856, National Archives, RG 121, entry 6.

91. I count forty-two office buildings (customhouse-post office-courthouse buildings); nine marine hospitals; and six other kinds of buildings, authorized in records of the Treasury. Some of the buildings may not have been started in the 1850s (National Archives, RG 121).

92. Quote from a letter to the president of Princeton College, in Constance Greiff, *John Notman*, 204. Cooper Hewitt & Company Collection in the Manuscript Division, Library of Congress. The collection contains business records of Trenton Iron Works, including a remarkable series of books from the years when the firm first manufactured beams. The books list customers and their orders, usually with the weight and price of the order. When the product was beams, the entry often included the name of the building for which they were destined. I tracked down most of the identified buildings; the great majority of these were entirely or partially fireproof.

93. Iron Business Shipping Record, September 9, 1852–March 25, 1854, Cooper Hewitt & Company Collection, Library of Congress, Manuscript Division. TIW shipped ninety-six "iron beams" to Peter Cooper during October and November 1853, which must have been compound beams. Peter Wight, who saw the build-

ing during renovations in the 1880s, confirms that the first floor contained compound beams (Peter Wight, *The Brickbuilder* 6 [March 1897]: 54). Esmond Shaw writes that they were used in three floors (*Peter Cooper and the Wrought Iron Beam*, 20). Photographs of the Harper & Brothers building being demolished show that some floor beams were double-channels.

94. Peter Wight, "Remarks on Fire-Proof Construction," *ARABJ* 2 (August 1869): 102. Wight read this essay, the first survey of American fireproof construction practice, at a meeting of the New York chapter of the AIA, and it was also published in the AIA, *Proceedings of the Second Annual Convention* (1868). Wight became an authority on fireproof construction.

95. Eugene Exman, *The Brothers Harper* (New York: Harper & Row, 1965).

96. Jacob Abbott, *The Harper Establishment; or, How the Story Books Are Made* (1855; Hamden, Conn.: Shoestring, 1956). Hewitt's responsibility for the fireproof floor chapter, Esmond Shaw, *Peter Cooper and the Wrought Iron Beam*, 18–19.

97. Montgomery Schuyler, "A Great American Architect: Leopold Eidlitz," *The Architectural Record* 24 (July–December 1908): 280. The iron contractor for the Mechanics Bank, James L. Jackson Brothers, received a big shipment of iron beams and four columns from Trenton in 1855. In 1856 and 1857, Trenton shipped iron beams, angle iron, pillars, and columns to the Continental Bank, and, in 1857, beams and box girders to the American Exchange Bank.

98. Quote from Peter Wight, "Remarks on Fire-Proof Construction," *ARABJ* 2 (August 1869): 102. Thomas R. Jackson was the architect and Stone and Witt the contractors of the Times Building. In 1857, Trenton shipped twenty-eight nine-inch beams, plus ten more beams of unknown size, to this project.

99. Astor Library: C & H Collection, Library of Congress, Cooper Hewitt daybook, shipment April 14, 1857. The original building opened in 1854 and an extension opened in 1859. Robert Jewett lists the Astor Library as a contender for the first building to have rolled beams ("Solving the Puzzle of the First American Structural Rail-Beam," 386). To clear up the matter, the first shipment of iron to the library, seventeen "trough bars," was sent in 1855, by which time iron beams were being used in government projects around the United States. S. B. Althause, who was the iron contractor for the Assay Office, built the Astor extension. Tatham & Brothers: R. G. Dun & Company Collection, Baker Library, Harvard Graduate School of Business Administration, New York, vol. 316, 87.

100. "Description of the Establishment of Cornelius & Baker," n.d. (1860?), in the collection of the Philadelphia Athenaeum; illustration in Charles Peterson, "Inventing the I-Beam" (1983).

101. 182 nine-inch beams were shipped from Trenton. John M. Gries designed the bank. Located at 421 Chestnut Street, the fireproof arches are visible in the first floor.

102. Information on the Kearsley Home, designed by John M. Gries, provided by Carl Baumert, structural engineer, November 3, 1998. On the Pennsylvania Railroad building, Edwin Betolett, "Fire-proof Construction in Philadelphia," *Insurance Engineering* 2 (1901); letter from Cooper Hewitt & Company to Samuel Reeves, Phoenix Iron Company, February 16, 1857 (Phoenix Steel Corporation, Hagley Library, Acc. 683, correspondence 1846–62). Located at 3rd Street and Willig's Alley, the building was designed by Stephen D. Button. Unfortunately, few of Phoenix's surviving records indicate who was buying the firm's beams in the 1850s and 1860s. This complicates tracing the history of fireproof construction, especially for cities like Philadelphia and Baltimore, where Phoenix might have done a bigger business than Trenton did.

103. Bliss Forbush, *The Sheppard & Enoch Pratt Hospital, 1853–1970* (Philadelphia: J. B. Lippincott, 1971), chapter 1; Phoenix Steel Corporation Collection, Hagley Library Acc. 683, correspondence 1864–75, letters from J. S. Norris to S. J. Reeves, June 17, 1864, and June 23, 1864.

104. The plans for the Traders' Bank called for three floors of iron beams and brick arches, as can be seen from a drawing reproduced in James O'Gorman, *On the Boards* (Philadelphia: University of Pennsylvania Press, 1989), 51. In 1858, Trenton shipped twenty-six nine-inch beams for the building. For the Keene Court House, Trenton sent Smith, Felton & Company sixty-one nine-inch beams in 1858.

105. Contract with George W. Smith & Horace Felton, Boston, for Iron Works and Blacksmiths Works and Materials for a Custom House in San Francisco, March 25, 1852, National Archives, RG 121, entry 26, San Francisco Custom House, box 1116.

106. In 1856 and 1857, Trenton shipped a large amount of iron to the Suffolk Bank, located at 60 State Street: 143 9-inch beams, 10 7-inch beams, 25 trough bars, and 4 girders. The unnamed building at Congress and State Streets probably was the Merchant's National Bank, which added a third floor at this time. Gridley J. F. Bryant designed the interior (*AABN* 7 [January 14, 1880]: 58). Trenton shipped 46 9-inch beams to this project in 1857. Information on the Registry of Deeds office, *The City Hall, Boston* (Boston: City Council, 1866), 50.

107. David Bigelow, *History of Prominent Mercantile and Manufacturing Firms in the United States* (Boston, 1857), 109.

108. Union College information from drawings available at Schaffer Library, Union College, Schenectady, New York; thanks to Cara Molyneaux, Head of Administrative Services. Trenton shipped sixteen nine-inch beams to William Harlow, Vassar's contractor, on April 28, 1862 (Elizabeth Daniels, *Main to Mudd* [Poughkeepsie: Vassar College, 1987], 12). James Renwick Jr. designed the Main Building, on which construction began in 1861.

109. Princeton College: Cooper Hewitt daybook, July and August 1855. In 1857,

Trenton shipped 82 9-inch beams to the Detroit Water Works, and the next year, shipped 92 9-inch beams to Chicago for its celebrated Water Works.

110. Marcus Whiffen and Frederick Koeper, *American Architecture 1607–1976* (Cambridge: MIT Press, 1981), 201; Daniel Bluestone, "Civic and Aesthetic Reserve: Ammi Burnham Young's 1850s Federal Customhouse Designs," *Winterthur Portfolio* 25 (Summer–Autumn 1990): 131–56.

111. See, for example, R. G. Hatfield, "Fireproof Floors for Banks, Insurance Companies, Office Buildings and Dwellings," February 1868, and Peter Wight, "Remarks on Fire-Proof Construction," a paper delivered at the AIA meeting and published in *Proceedings of the Second Annual Convention* as well as in *ARABJ* 2 (August 1869).

112. *New York Times*, May 16, 1868, 2:5.

Chapter 3
Response to the Great Fires

American Institute of Architects, *Proceedings, Eleventh Annual Convention*, October 1877, p. 25.

1. R. G. Hatfield, "Fireproof Floors for Banks, Insurance Companies, Office Buildings and Dwellings," AIA *Proceedings of Second Annual Convention* (1868): 8.

2. Trenton Iron Company supplied a large number of beams for the courthouse in 1862. On the Women's Home, *History of Real Estate, Building and Architecture in New York City* (1898; New York: Arno Press, 1967), 459–61, 616, and *The Architectural Record* 1 (1891–92): 245. John Kellum designed both the courthouse and the Women's Home. Information on the banks from Peter Wight, "Remarks on Fire-Proof Construction," *ARABJ* 2 (August 1869): 102; *History of Real Estate... in New York City*, 379, 615; and *AABN* 5 (March 1, 1879): 166. Griffith Thomas designed the Park Bank; Robert G. Hatfield designed the City Bank; John B. Snook designed St. John's Park freight depot.

3. Trenton shipped beams to these two projects. Gamaliel King and Herman Technitz designed the courthouse.

4. On the Chicago Historical Society building, Paul Angle, *The Chicago Historical Society 1856–1956* (New York: Rand McNally, 1956); building designed by Burling & Company. Joliet prison, designed by William Boyington and Otis L. Wheelock, received a large quantity of beams from Trenton in 1861. Cleveland City Hall, *ARABJ* 2 (1870): 699; Heard and Blythe won the design competition for this building. On the Smithsonian, Adolf Cluss, "Report of the Architects . . . ," *Annual Report of the Board of Regents of the Smithsonian Institution* for 1884 (Washington, D.C.: GPO, 1885), and Robert Esau, "Fear of Fire; An Investigation into the Fireproofing of the Smithsonian Institution Building," Smithsonian Institution, August 10, 1990. Boston City Hall, *The City Hall, Boston* (Boston City

Council, 1866); Gridley J. F. Bryant and Arthur Gilman designed it. The White store was not standard fireproof construction; above the first floors, the brick arches, rather than being filled, functioned as barriers under heavy plank floors. Floors described in *ARABJ* 1 (August 1868): 90–94.

5. Ann Kelchburg with Ronald Mullins, *A History of the Continental Insurance Company* (New York: The Continental Corporation, 1979), 14. Griffith Thomas designed the building.

6. New York Life described in Asher and Adams, *Pictorial Album of American Industry 1876* (1876; New York: Routledge, 1976), 35; designed by Griffith Thomas. John Kellum designed the Mutual Life Insurance building; Frederic Diaper designed the Columbia Insurance building.

7. *The Nation* (April 11, 1867): 298.

8. Peter Wight, *The Brickbuilder* 6 (March 1897): 54. Buildings designed to have stone slab floors include the War and Navy Offices in Washington (1844) and a fireproof private library in Paterson, New Jersey (House report no. 267, 28th Congress, 1st session, 1844, appendix A, and "Fireproof Library," *JFI* 12, 3rd series [November 1846]: 337–38). On the buildings with stone floors, Peter Wight, "Remarks on Fire-Proof Construction," 104. Leopold Eidlitz designed the American Exchange Bank and Continental Bank. The typical British version of the floor differed from the American version in that in Britain, stone slabs usually were put on top of the beam, while Americans put the stones on the bottom flange of the beam.

9. In David Bigelow, *History of Prominent Mercantile and Manufacturing Firms* (Boston, 1857) and reproduced in John Kouwenhoven, *The Columbia Historical Portrait of New York* (New York: Harper & Row, 1972), 245. The ad stated, "The Proprietors for many years have given much attention to the construction of Fireproof Buildings, and by practical experience (of which they have had much), they are fully prepared to give such information as will be of great importance to parties desiring to put up such buildings. Persons out of the city desiring to purchase IRON WORK, by sending plans of their buildings, or a description of what they want, will be promptly attended to, and furnished with drawings and estimates (if desired) of the cost of the articles they desire to purchase" (103). John B. Cornell held several patents, including one for a wooden fireproof floor (no. 22,939, February 15, 1859), but not for this floor.

10. On the Bank of the State of New York, Francis Kowsky, *Country, Park, and City: The Architecture and Life of Calvert Vaux* (New York: Oxford University Press, 1998), 82–85. Calvert Vaux and Frederick Withers designed the bank. Details of the building in *Engineering News* 50 (September 10, 1903): 214–15. On other buildings with these floors, Peter Wight, "Remarks on Fire-Proof Construction," AIA *Proceedings* (December 1868): 61 and error sheet, and *AABN* 1 (April 15, 1876): 126. Peter Wight wrote that floors made of corrugated plates

were used in "some instances," which suggests they were not used much (*The Brickbuilder* 6 [March 1897]: 54).

11. Treasury Department, National Archives, RG 121, entry 13, Letters Sent, 1855–63; Office of the Curator, Department of the Treasury, Diagram of Corrugated Iron dated July 25, 1863; personal inspection, December 10, 1996.

12. Patent no. 64,659, dated May 14, 1867; *ARABJ* 1 (July 1868): 60–67.

13. William Fairbairn, *On the Application of Cast and Wrought Iron to Building Purposes* (New York: John Wiley, 1854), 140–45. An 1868 article about Gilbert's floor quotes Fairbairn on the merits of this kind of floor, so Gilbert knew Fairbairn's book (*ARABJ* 1 [July 1868]: 65–66). Fairbairn spaced his sixteen-inch-deep beams ten feet apart. Cornell and Gilbert both spaced their beams six feet apart. This similarity may be a coincidence or may show that Gilbert knew about Cornell's floor (*Engineering News* 50 [September 10, 1903]: 222, and *ARABJ* 1 [July 1868]: 67). Corrugated iron arches eventually were used in England, in a system introduced there by Richard Moreland in the 1860s (John J. Webster, "Fireproof Construction," Institution of Civil Engineers *Minutes of Proceedings* 12 [1890–91]: 264). Moreland patented his corrugated iron floor in 1867, no. 885 of 1867. Thanks to Lawrance Hurst for the date of the patent.

14. *ARABJ* 1 (July 1868): 67. The Franklin Institute committee described the fill as "cement," which would have been a stronger material than ordinary mortar-based concrete.

15. *Chicago Tribune*, May 28, 1868, quoted in *ARABJ* 1 (July 1868). *AABN* 1 (April 15, 1876): 126. Edward Burling designed both the First National Bank and the Tribune buildings. Source for the wings of the courthouse and Fidelity Safe Deposit Company building, George Dwight Jr. in *AABN* 1 (July 8, 1976): 222–23. Although other sources say the Nixon had brick floors, Henry Ericsson wrote that it had corrugated iron ceilings, "the same as the Tribune Building" (*Sixty Years a Builder* [1942; New York: Arno Press, 1972], 206). Otto Matz designed the Nixon. On the new post office, *AABN* 56 (June 5, 1897): 73.

16. Peter Wight, *AABN* 1 (August 5, 1876): 255.

17. George Dwight obituary, *The Springfield Republican*, January 3, 1909. (Thanks to Maggie Humberston of the Connecticut Valley Historical Museum.) Another source described the building as having brick floors so it is uncertain where the iron arches were used (*A History of Real Estate . . . in New York City*, 379–82). Arthur D. Gilman and Edward H. Kendall designed the Equitable in New York with George B. Post as consulting architect. Exactly how the iron arches were used in the building—whether in the floors or roof—is unknown. On the Equitable in Boston, information was obtained from John E. Whiting, *A Schedule of Buildings . . . in the City of Boston* (published for the use of insurance companies, 1877).

18. G. K. Gilbert et al., *The San Francisco Earthquake and Fire of April 18, 1906 and Their Effects on Structures and Structural Materials*, Bulletin no. 324, Department of the Interior, USGS (Washington, D.C.: GPO, 1907), 36. The structure was built in stages over a thirty-year period, so it is not possible to know when the iron arches were installed.

19. Patent no. 12,642, April 3, 1855.

20. The tile contained fire clay (fire clay is a kind of clay used to make refractory brick). Peter Wight, an experienced tile manufacturer, visited the building when it was being renovated and saw the tiles (*The Brickbuilder* 6 [March 1897]: 54).

21. James Bowen, "Report upon Building, Building Materials and Methods of Building," *Reports of the United States Commissioners to the Paris Universal Exposition, 1867*, vol. IV (Washington, D.C.: GPO, 1870), 32.

22. Abram Hewitt, "The Production of Iron and Steel in Its Economic and Social Relations," Paris Universal Exposition, 1867, Reports of the United States Commissioners (New York, 1870), 7.

23. Recommendations for wood construction from *Building News* 12 (January 27, 1865): 69, and in, for example, E. M. Shaw, *Fire Surveys* (London: Effingham Wilson, 1872), 43. Opposition to stone, *Building News* 22 (March 29, 1872): 250. *Fire Surveys* went through several editions.

24. For example, B. H. Thwaite, *Our Factories, Workshops, and Warehouses* (London: E. & F. N. Spon, 1882), 207.

25. *Building News* 12 (April 7, 1865): 244.

26. *Building News* 22 (March 29, 1872): 250. The first British-made covering for iron girders—a tile casing—dates from 1873. The architect John Whichcord, who used it in the National Safe Deposit Company building, introduced it (B. H. Thwaite, *Our Factories*, 210).

27. "Fire-Proof Construction," *ARABJ* 2 (March 1870): 500.

28. *Architects' and Mechanics' Journal* 2 (May 5, 1860): 42.

29. According to Peter Wight, an early authority on fireproof construction, "Before 1871 there were few occasions of disastrous fires to prove the inefficacy of . . . simply substituting iron for wood." He knew of only two fireproof buildings that had burned: the Singer Sewing-Machine Company building and the Fulton Bank, both in New York City (Peter Wight, "Fireproof Construction and the Practice of American Architects," *AABN* 41 [August 19, 1893]: 113).

30. John Neal, *Account of the Great Conflagration in Portland, July 4th & 5th, 1866* (Portland: Starbird & Twitchell, 1866), 5, 7, and 26.

31. While the damage to its granite facade could have been repaired, the supervising architect, Alfred Mullett, preferred to replace it with one of his signature French Second Empire buildings. Incredibly, despite ample evidence that fire

readily damaged granite, Mullett used granite to build the new customhouse (1868–71) (*AABN* 7 [January 31, 1880]: 33).

32. Richard Levy, *The Professionalization of American Architects and Civil Engineers, 1865–1917* (Ph.D. diss., University of California–Berkeley, 1980), chapter 2. The AIA began to hold annual conventions in 1867 (A. J. Bloor, "History of American Institute of Architects," *IA* 16 [October 1890]: 40–42).

33. *The Crayon*, an arts magazine, carried news of, and published some papers given at, the meetings of the AIA, but it ceased publication during the Civil War. Likewise, the short-lived *Architectural Review and American Builders' Journal* stopped during the war. Neither resumed publication after the war.

34. About nine inches of rain was normal. Elias Colbert and Everett Chamberlin, *Chicago and the Great Conflagration* (Cincinnati and New York: C. F. Vent, 1872); Arthur Cote, ed., *Fire Protection Handbook*, 18th ed. (Quincy, Mass.: NFPA, 1997), 1–8.

35. Elias Colbert and Everett Chamberlin, *Chicago and the Great Conflagration*, chapter 2.

36. James W. Sheahan and George Upton, *The Great Conflagration* (Chicago: Union Publishing Company, 1872), 77.

37. Part of the courthouse survived and, after repairs, continued to be used for a time (*AABN* 1 [April 15, 1876]: 126). Source on the Fidelity Safe Deposit Company building: George Dwight Jr. in *AABN* 1 (July 8, 1976): 222–23. Dwight wrote that a Newfoundland dog locked inside this building survived the fire. Likewise, the safe and vaults of First National Bank were unharmed, and the bank resumed operations so quickly after the fire (it was "rebuilt in December" of 1871) that part of the original structure must have been reused (Frank Randall, *History of the Development of Building Construction in Chicago* [Urbana: University of Illinois Press, 1949], 48). On the Tribune Building, Horace White, editor of the *Tribune*, quoted in James W. Sheahan, *The Great Conflagration*, 227–28.

38. Peter Wight on the fire at Chicago, AIA *Proceedings of the Fifth Annual Convention* (1871): 47.

39. Quotes from *Real Estate Record* 10 (November 16, 1872): 185.

40. *Report of the Inspector of Buildings*, City Doc. No. 12, City of Boston Documents, 1872.

41. Edward Atkinson's estimate, in "The Protection of City Warehouses from Loss by Fire," February 1, 1883.

42. AIA, *Proceedings of the Seventh Annual Convention* (1873): 18–19. James H. Giles designed the Lord & Taylor store (*A History of Real Estate ... in New York*, 616). The American Express Company building was the work of Charles D. Gambrill and Henry H. Richardson (Frank Randall, *History of the Development of Building Construction in Chicago*, 80). Peter Wight worked on both buildings, as

construction superintendent on Lord & Taylor and as consulting architect on the American Express building.

43. AIA, *Proceedings of the Seventh Annual Convention* (1873).

44. A. J. Bloor, "Report on Mansard Roofs," AIA, *Proceedings of the Seventh Annual Convention* (1873): 40–48; Robert Hatfield, "Anti-Fire Construction," *AABN* 2 (December 15, 1877): 399–400; Detlef Lienau, "Fire-Proof Construction," *AABN* 3 (January 5, 1878): 5–6; F. Schumann, "Fire-Proof Construction," *AABN* 3 (June 24, 1878): 224–25 and 4 (July 6, 1878): 4–5; Peter Wight, "The Fire Question," 3 *AABN* (March 2, 1878): 75–76.

45. Robert Hatfield, comments on the Boston, AIA, *Proceedings of the Seventh Annual Convention* (1873): 19. In tests, only sandstone—the brownstone of New York City—proved able to withstand fire and water well.

46. L. Eidlitz, "Cast Iron and Architecture," *The Crayon* 6 (January 1859): 21.

47. A. J. Bloor, "Report on Mansard Roofs," AIA *Proceedings* (1873): 41.

48. Dorothy Stroud, *Sir John Soane Architect* (London: Faber & Faber, 1984), 156. Alan Menuge, "The Cotton Mills of the Derbyshire Derwent and Its Tributaries," *Industrial Archaeology Review* 16 (Autumn 1993): 43 and 48. John J. Webster, "Fire-proof Construction," 264–65.

49. S. B. Hamilton, *A Short History of the Structural Fire Protection of Buildings,* National Building Studies, Special Report no. 27 (London: HMSO, 1958), 15; Henry Eyton, "On the Fire-proof Construction of Dwellings," *Journal of the Society of Arts* 12 (September 20, 1864): 721–22; Henry Roberts, "Dwellings of the Labouring Classes," *Civil Engineer and Architect's Journal* 13 (April 1850): 123–27.

50. William Fairbairn, *On the Application of Cast and Wrought Iron to Building Purposes,* 134; construction dates, Colum Giles and Ian Goodall, *Yorkshire Textile Mills, 1770–1930* (London: HMSO, 1992), 102.

51. Henry Eyton, *Journal of the Society of Arts* 12 (September 20, 1864): 722; John J. Webster, "Fire-proof Construction," 269. The Bunnett arch could span twenty-one feet; the arch in Saltaire mill spanned ten feet (J. J. Webster, "Fire-proof Construction," 264 and 269).

52. Charles Eck, *Traité de Construction en Poteries et Fer* (Paris: J. C. Blosse, 1836). A practical mason, who reported on the construction of houses in Paris for the U.S. commissioners to the Paris Universal Exposition in 1867, described the floors as consisting of arched I-section beams, about two feet apart, filled with hollow bricks bedded in plaster (James Bowen, "Report upon Building, Building Materials and Methods of Building," 34).

53. William J. Fryer Jr. asserted that the first use of flat hollow tile arches in the United States (or anywhere) was in the New York Post Office (*The Architectural Record* 1 [1891–92]: 230). It is more likely that the Kendall Building in Chicago was the first.

54. Patent no. 112,926, March 21, 1871.

55. A. T. Andreas, *History of Chicago,* vol. 3 (Chicago: A. T. Andreas, 1886), 87–88. Johnson received ten patents in 1869 for items connected with grain elevators (*Annual Report of the Commissioner of Patents* for the year 1869).

56. Patent no. 112,930. Kreischer's company, in 1902 styled B. Kreischer and Son, New York Fire Brick, was established in 1845. Regarding the Assay Office, "Abstract of Disbursements for Erection of Assay Office," September 30, 1854, National Archives, RG 121, entry 26, Assay Office, box 285.

57. Peter Wight, *The Brickbuilder* 6 (March 1897): 54.

58. Partition blocks, patent no. 112,925, March 21, 1871. Floor blocks, patent 132,292, October 15, 1872, with W. Freeborn. On French systems known in the United States, Peter Wight, *The Brickbuilder* 6 (March 1897): 55.

59. *Industrial Chicago* (Chicago: Goodspeed, 1891), 403.

60. On the courthouse and the Palmer House, Henry Ericsson, *Sixty Years a Builder,* 206; A. T. Andreas, *History of Chicago,* vol. 3, 242; and Frank Randall, *History . . . of Building Construction in Chicago,* 54. Van Osdel designed the Palmer House. James J. Egan designed the courthouse and jail. Buildings fireproofed by the Pioneer Fireproof Construction Company, successor to Johnson & Company, were listed in *IA* 4 supplement (August 1884).

61. Peter Wight, *The Brickbuilder* 6 (April 1897): 74. In a letter to Mullett dated May 26, 1873, HHC wrote that it held patents covering "all forms of sectional hollow tiles used in the construction of arches" (National Archives, RG 121, entry 26, New York Post Office, box 312). Neither Heuvelman nor Haven had relevant patents in their own names.

62. R. G. Dun & Company Collection, Baker Library, Harvard University Graduate School of Business Administration, New York, vol. 124, 285, 226, 226W, 238 and vol. 373, 1430. An ad for Novelty Iron Works in 1868 read, "they are prepared to furnish complete fire-proof structures, fronts, columns, lintels, floors, roofs, etc." (Henry T. Brown, *Mechanical Movements* [New York, 1868]).

63. Quotes, letter from HHC to A. B. Mullett, March 15, 1873, National Archives, RG 121, entry 26, New York Post Office, box 312. On the use of the tiles, construction superintendent Calvin Hulburd to A. B. Mullett, October 9, 1873, National Archives, RG 121, entry 26, New York Post Office, box 312. The tiles were installed in the halls and the large rooms on the third and fourth stories on the Park front, and the fifth floor. They were described in a letter as being about six inches wide by twelve inches long. J. K. Freitag, an authority on fireproof construction, wrote that Leonard Beckwith made the arches in this building (J. K. Freitag, *The Fireproofing of Steel Buildings* [New York: John Wiley & Sons, 1899], 11). However, the Treasury Department records indicate that HHC made them; I found no mention of Beckwith in the records.

64. Leonard F. Beckwith, "Report on Béton-Coignet, its Fabrication and Uses," *Reports of the United States Commissioners to the Paris Universal Exposition, 1867,* vol. 4 (Washington, D.C.: GPO, 1870). This is available in the serial set, no. 1354. Francois Coignet invented the material, hence the name. On Beckwith's education, Peter Wight, "Fireproof Construction and the Practice of American Architects," 113. Information on the company was from R. G. Dun & Company Collection, Baker Library, Harvard GSBA, vol. 326.

65. Patent no. 112,929, March 21, 1871. Peter Wight writes that Beckwith's company "control[led] certain process patents for using cement and plaster" (*The Brickbuilder* 6 [April 1897]: 74). A search of the United States Patent Office official gazettes for Leonard Beckwith patents, using the pre-1873 subject index and annual indexes through the mid-1870s, turns up no patents for concrete.

66. Patent no. 151,826 (May 9, 1874). Arthur also wrote a report for the U.S. commissioners to the Paris Universal Exposition of 1867 on asphalt and bitumen for making streets and sidewalks. R. G. Dun records list Arthur as a consulting architect to the company in 1873; by 1884, he was a superintendent (advertisement for the company, *AABN*, July 12, 1884). On the firm's license with Johnson, Peter Wight, "Fireproof Construction and the Practice of American Architects," 114.

67. On the Natural History Museum, Francis Kowsky, *Country, Park, and City,* 235. Calvert Vaux and Jacob Wrey Mould designed the building. Edmund G. Lind designed the Peabody Library. A partition block salvaged from a recent renovation is made of concrete.

68. Perhaps both materials were used to make arches in the Tribune Building, which would explain the disagreement between sources noticed by the historian Sarah Landau. She writes that one source—the *New York Daily Tribune* itself—described the floors as cement blocks while *New-York Sketch-book* of 1874 described the arches as terra cotta ("Richard Morris Hunt: Architectural Innovator and Father of a 'Distinctive' American School," in Susan Stein, ed., *The Architecture of Richard Morris Hunt* [Chicago: University of Chicago Press, 1986], 58 and 74, footnote 30). Possibly the clay arches were stronger than the concrete ones, so FBC put them in places that were to be finished with heavy, mosaic tiles. Wight wrote that "the avowed object at the time was that they [clay arches] would be better than cement [blocks] in hallways and rooms which were to be finished with encaustic tile floors" (Peter Wight, *The Brickbuilder* 6 [April 1897]: 74). Yet the concrete block floors in the American Museum of Natural History were paved with mosaic tiles.

69. Peter Wight, "Fireproof Construction and the Practice of American Architects," 114.

70. R. M. Hunt, "The Architectural Exhibits of the International Exhibition," *Reports of the U.S. Centennial Commission* (Washington, D.C.: GPO, 1880), 14 and 71. Hunt designed the Coal and Iron Exchange and Tribune buildings and there-

fore was one of Beckwith's first customers (advertisement in *AABN* 3 [January 5, 1878]). It would also be interesting to know how FBC manufactured its various products. Unfortunately, no trade catalogue for FBC, which might shed light on the its operations, has turned up.

71. On the Illinois state capitol and Singer Building in St. Louis, Peter Wight, *The Brickbuilder* 6 (April 1897): 74 and *IA* 5, extra no. (April 1885): 46.

72. Quotes from *The New Southern Hotel*, 1881, 5. The *Commercial and Architectural St. Louis* (Jones & Orear, 1888) described the interior as being constructed of "wrought iron with fire-proof blocks between." But a brochure about the hotel says the floorboards were "laid on solid cement" (*The New Southern Hotel*, 1881). Peter Wight described the floors as being made of "iron rails and concrete," which suggests a slab (*The Brickbuilder* 6 [April 1897]: 75). George I. Barnett, the building's architect, worked on the Illinois state capitol with Alfred Piquenard and would have seen the fireproof blocks in this building. Perhaps the floors were made of blocks between iron rails, leveled up with concrete.

73. *AABN* 3 (February 2, 1878): 43; *AABN* 5 (May 31, 1879): 172. E. Townsend Mix designed the building.

74. In an 1878 advertisement, for example, FBC offered "Hollow Blocks of any pattern," which suggests that blocks could be made to order (*AABN* 3 [January 5, 1878]). Peter Wight believed that hollow tile manufacturers outside Chicago used "any refuse clay" in making arch blocks (*The Brickbuilder* 6 [April 1897]: 75). But documentary evidence indicates otherwise. For example, in trying to persuade Mullett to use its hollow tile, HHC wrote that it had "found a superior quality of clay of which to manufacture them," suggesting that the company was not indifferent to the material (HHC to Mullett, March 15, 1873, National Archives, RG 121, entry 26, New York Post Office, box 312). Beckwith made his blocks of a variety of costly materials, including fire clay. Johnson & Company's arches for Chicago City Hall were made of fire clay, although the courthouse arches were not (Peter Wight, *The Brickbuilder* 6 [April 1897]: 74–75).

75. Peter Wight, *AABN* 5 (May 31, 1879): 172.

76. HHC to A. B. Mullett, May 26, 1873, National Archives, RG 121, entry 26, New York Post Office, box 312.

77. Advertisement in *AABN* 3 (January 5, 1878).

78. *Terra Cotta Lumber* (New York, 1882), 29–33.

79. When the Civil War began, the engineer-in-charge at the Treasury Department, Capt. William Franklin, left to fight, and for some unknown reason, his position disappeared. From about 1866 to 1874, Alfred B. Mullett held the supervising architect job and he reversed his predecessors' design priorities: rather than technologically advanced buildings with simple facades, he produced fashionable French Second Empire-style buildings that had what were, by this time, techno-

logically conventional structures. While Mullett did adopt some of the new materials that came on the market during his tenure, he did not use government procurement as a means to encourage innovation. A complaint about this situation is found in "Civis," "The Office of the Supervising Architect; What it was, what it is, and what it ought to be," New York, 1869 (collection of the AIA archives). "Civis" probably was Adolf Cluss.

80. Peter Wight, *The Brickbuilder* 6 (April 1897): 74.

81. Peter Wight, "Fireproof Construction and the Practice of American Architects," 114.

82. Henry Oakley, "Address," Decennial Meeting, NBFU of the U.S., April 26, 1876. Projects listed in a Fire-proof Building Company advertisement, *AABN* 3 (January 5, 1878). The Bennett and Drexel buildings were designed by Arthur Gilman; the Western Union by George B. Post; and Coal & Iron Exchange, Tribune, and Lenox Library by Richard M. Hunt.

83. Patent no. 140,352, December 3, 1872.

84. Patent no. 151,826, May 9, 1874. Peter Wight wrote that the first patent granted to an American for an end-construction arch went to Levi Scofield (*The Brickbuilder* 6 [May 1897]: 99.) However, Scofield's patent (patent no. 161,357, December 12, 1874) postdated Beckwith's and his design—a large, arched tile with small cavities that spanned between beams—would have been difficult to make. In the 1850s, Joseph Bunnett patented an end-construction style arch; consequently, Beckwith's was not the very first example.

85. Patent no. 143,351, July 14, 1874; William Fryer Jr., *Architectural Iron Work* (New York: John Wiley & Sons, 1899), 129.

86. Patent no. 221,501, filed October 3, 1879; *The Brickbuilder* 6 (May 1897): 99.

87. Patent no. 143,352, August 12, 1873.

88. Chicago Water Works roofs, Peter Wight, *The Brickbuilder* 6 (April 1897): 75. Methods of building roofs at the close of the 1870s, see Peter Wight, *AABN* 5 (May 31, 1879): 172.

89. Patent no. 156,361, October 27, 1874. *AABN* 1 (December 30, 1876): 421. Little information about Loring's life has turned up, not even the dates of his birth and death. His only published work, that I have found, was a book of building plans made with his partner at the time, William Le Baron Jenney (*Principles and Practice of Architecture* [Chicago, 1869]). Sharon Darling, in her *Chicago Ceramics and Glass, an Illustrated History from 1871 to 1933* (Chicago: Chicago Historical Society, 1979), gives some information about Loring's terra cotta business, but does not cite the sources of this information. The Chicago Historical Society had no further information about the man.

90. For information on the architectural terra cotta business in Chicago, see

Sharon Darling, *Decorative and Architectural Arts in Chicago, 1871–1933* (Chicago: University of Chicago Press, 1982).

91. For example, Peter Wight in *AABN* 1 (January 22, 1876): 30–31.

92. It continued to be used until at least 1895, e.g., for column protection in the Siegel-Cooper store in New York City; "Burnt Clay Fireproofing and Its Substitutes," *Architectural Record* 8 (July–September 1898): 112.

93. New York laws of 1871, ch. 625, section 14. The architect and engineer Nathaniel H. Hutton recommended columns that were concentric tubes of iron with the space between filled with plaster or cement ("Fire-proof Construction," *Scientific American* supplement no. 10 [March 4, 1876]: 158). Peter Wight reported that in unheated buildings in cold climates, the plaster in these double columns was liable to freeze and break the outer shell, so builders often left out the plaster (*IA* 19 [July 1892]: 71). J. B. Cornell, the New York iron founder, held a patent on a "double cylinder" or "fire-proof" column similar to this, consisting of an inner cast iron column and outer cast iron shell (patent no. 27,528, March 20, 1860).

94. Patent no. 154,852, August 10, 1874; Peter Wight, "Recent Fireproof Building in Chicago," Part VI, *IA* 19 (July 1892): 71; Peter Wight, *The Brickbuilder* 6 (August 1897): 173.

95. *AABN* 3 (March 23, 1878): 102–3; *AABN* 1 (August 19, 1876): 271. The building was one of the earliest to use architectural terra cotta, including molded tiles and enameled brick, manufactured by Chicago Terra Cotta Works. Treat and Foltz designed the building.

96. *AABN* 5 (May 31, 1979): 172. On the buildings containing his columns, Peter Wight, "Recent Fireproof Building in Chicago," Part VI, *IA* 19 (July 1892): 72.

97. 1874 patent no. 154,852, and 1878 patent nos. 203,972 and 204,867. Peter Wight, "Recent Fireproof Building in Chicago," Part VI, *IA* 19 (July 1892): 72.

98. Peter Wight, "Recent Fireproof Building in Chicago," Part VI, *IA* 19 (July 1892): 72.

99. Patent no. 248,094, October 1, 1881; mentioned in *IA* 4 (September 1884): 27. In 1885, Gilman filed for patents for a variety of applications for his terra cotta lumber: patents no. 338,509–338,518. On the success of the material, U.S. Geological Survey, "Mineral Resources of the U.S., Calendar Year 1885," (GPO, 1886), 422, 573–74.

100. Peter Wight, "Recent Fireproof Buildings in Chicago," Part VI, *IA* 19 (July 1892): 69; Wight, *The Brickbuilder* 6 (April 1897): 75.

101. Report on bids for ironwork for the basement and subbasement, Office of the Supervising Architect, June 9, 1870, National Archives, RG 121, entry 26, New York City Post Office, box 307.

102. Names from *Terra Cotta Lumber* (New York, 1882).

103. *New York Times,* May 16, 1868, 2:5. Richard Hoe's patent, for a "fire-proof building" no. 90,361, May 25, 1969.

104. Peter Wight, *AABN* 5 (May 31, 1879): 171 and (June 7, 1879): 179.

105. Wire net had been used in England before the 1840s, in the Pantechnicon; a style of wire net was patented in 1841 (J. J. Webster, "Fire-proof Construction," 274). Expanded metal was an American product, introduced around 1884 (*Iron Age* 42 [July 5, 1888]: 34).

106. *IA* 3 (July 1884): 84.

107. Patent no. 145,211, December 2, 1873.

108. *AABN* 5 (June 7, 1879): 179.

109. Patent no. 133,448, November 26, 1872.

110. Patent no. 143,196 and 143,197, September 23, 1973. Tartiere's position from R. G. Dun & Company Collection, Baker Library, Harvard GSBA, New York, vol. 326, 1180.

111. Patent no. 156,808, October 21, 1874. Information on Maurer from R. G. Dun & Company Collection, Baker Library, Harvard GSBA, New York, vol. 379, 147, 149, 187, and 200 a/52.

112. Patent no. 202,617, December 21, 1877.

113. Speech reprinted in *AABN* 16 (November 15, 1884): 232. I have found a few buildings with brick floors that date from the latter part of the century in Boston. One of these is Faneuil Hall: in the late 1890s, in order to safeguard this historically significant building from fire, the city rebuilt the first floor over the food market using steel beams and brick arches.

114. Peter Wight, "Recent Fireproof Building in Chicago," *IA* 5, extra no. (April 1885): 52–53.

115. Edwin Betolett, "Fireproof Construction in Philadelphia," *Insurance Engineering* 2 (1901). Addison Hutton designed this building.

116. The Patent Office and Smithsonian rebuilding projects were prompted by a fire that destroyed the north and west wings of the Patent Office in 1877. Cluss worked on the Castle in 1866, when he supervised the reconstruction of its fire-damaged roof. Denis R. McNamara, *"One of Our Best and Most Talented Men": Adolf Cluss and German Theory in Nineteenth Century Washington, D.C.* (M.A. thesis, School of Architecture, University of Virginia, 1994), 7. "Report of the Architects for the Reconstruction of the Eastern Portion of the Smithsonian Institution," *Annual Report of the Board of Regents* for 1884 (Washington, D.C.: GPO), xxi. Peter Wight fireproofed the Pension Building's columns. *IA* 3 (July 1884): 82.

117. Examples of Wight's semifireproof projects include the Ryerson/Revell

Building in Chicago (Adler and Sullivan, 1881–83), the American Bank Note Company Building (c. 1884) in New York, and the Tribune Building (Leroy Buffington, c. 1884) in Minneapolis. Pioneer's Chicago projects included the Calumet Building (c. 1882–84), which had tile partitions in addition to tile-protected floors, and its own office building (c. 1884). For a list of the firm's projects, see *IA* 5 (July 1885): 3. For a description of Pioneer's own building, see *IA* 4 (August 1884): 12.

118. Quote, *AABN* 6 (August 16, 1879): 55. HHC ceased making hollow tile but continued to make architectural iron; by 1878, Haven had left, and the firm became Heuvelman & Company (Dun Collection, New York, vol. 373). After Johnson's death in 1879, his son Ernest joined the firm, which was renamed Ottawa Tile Company after the town where its clay mines and works were located (A. T. Andreas, *History of Chicago*, 87–88). Around 1884, the company changed its name to Pioneer (*IA* 3 [February 1884]: 8 and *IA* 4 [August 1884], advertising supplement). On the Fire-proof Building Company, *IA* 5 (July 1885): 3. On the Raritan Hollow and Porous Brick Company, Advertisers' Trade Supplement, *AABN* 24 (December 1, 1888). New York Terra Cotta Lumber Company, for one, made tile for Wight (*New York Terra Cotta*, 1882, 34). On John J. Schillinger's firm, advertisement in *AABN*, e.g., November 15, 1884, xix. Schillinger's patents: nos. 213,945 and 221,108 (1879); 233,029 (1880); 252,263, 262,483, 262,284 (1882); 293,523 (1884).

119. Although wrought iron melts at a much higher temperature than cast iron, it begins to weaken at a lower temperature. In the 1870s, even people who should have known better neglected to protect beam flanges. For example, William J. Fryer, in his textbook on architectural ironwork, wrote that owners preferred to "leave exposed the lower flanges of the iron beams." Rather than warning them against doing so, however, Fryer recommended "moulding" the bottom flanges to make them more attractive (*Architectural Iron Work*, 125).

120. Peter Wight, *The Brickbuilder* 6 (May 1897): 98. Maurer's patent, no. 299,820, December 5, 1883.

121. Peter Wight, *The Brickbuilder* 6 (May 1897): 98. Wight's patent, no. 285,452, filed March 29, 1883.

122. *Annual Report of the Board of Regents for the Year 1866* (Washington, D.C.: GPO, 1867), 69; National Archives, RG 121, entry 26, New York City Post Office, box 311.

123. Massachusetts laws of 1872, ch. 371, sections 19 and 20; New York laws of 1871, ch. 625, section 13.

124. *AABN* 5 (May 31, 1879): 171.

125. John R. Freeman, "Fire Protection Engineering," lecture notes, c. 1894, JRF papers, MIT Archives, box 29.

126. John R. Freeman, "Fire Protection Engineering."

Chapter 4
Mill Fire Protection Methods
Enter the Mainstream

The Fire-Engineer, the Architect, and the Underwriter: Their Relations to Each Other (Boston: Franklin Press; Rand, Avery, & Co., 1880), p. 17.

1. Keith A. Falconer, "Fireproof Mills—The Widening Perspectives," *Industrial Archaeology Review* 16 (Autumn 1993): 15–16.

2. Charles J. H. Woodbury, "Methods of Reducing Fire Loss," *Cassier's Magazine* 1 (November 1891–April 1892): 29.

3. The most detailed history of mill architecture is Betsy Bahr, *New England Mill Engineering: Rationalization and Reform in Textile Mill Design, 1790–1920* (Ph.D. diss., University of Delaware, 1987). Another history of AFM and its influence on mill design is John J. Crnkovich, "The New England Mutuals' Influence on Industrial Architecture" (January 1978, unpublished; copy at the National Museum of American History). For a history of the AFM organization, which contains a chapter on mill construction, see *The Factory Mutuals 1835–1935* (Providence: Manufacturers Mutual Fire Insurance Company, 1935).

4. J. R. Freeman, "Fire Protection Engineering," lecture notes, JRF papers, MIT Archives, box 29.

5. In his history of BMMFIC, Dane Yorke gives Atkinson principal credit for publicizing slow-burning construction. Yorke had access to company records that apparently no longer exist (BMMFIC is defunct) (*Able Men of Boston . . . the . . . story of . . . the Boston Manufacturers Mutual Fire Insurance Company* [Boston: BMMFIC, 1950]). In contrast, Atkinson's biographer Harold Williamson credits him with having had "considerable" influence "in securing a more widespread adoption of safer building construction and better fire protection," but not with playing a key role (133) (*Edward Atkinson, the Biography of an American Liberal 1827–1905* [Boston: Old Corner Bookstore, 1934]). From Atkinson's personal papers and the statements of contemporaries, I side with Yorke. Atkinson wrote he coined the term "slow-burning construction" and I have found nothing to contradict this (*AABN* 40 [April 1, 1893]: 14).

6. Dane Yorke, *Able Men*, 121, schedule reproduced from Arthur Ducat, *The Practice of Fire Underwriting*.

7. Nicholas Wainwright, *The Philadelphia Story: The Philadelphia Contributionship . . .* (Philadelphia: The Philadelphia Contributionship, 1952), 61–62, 91; *The Fire Association, A Short Account of the Origin and Development of Fire Insurance in Philadelphia* (Boston: The Fire Association, 1917).

8. Dane Yorke concluded that "the real novelty of Zachariah Allen's approach to mutual fire insurance was that he planned his company for manufacturers only" (Dane Yorke, *Able Men*, 30).

9. Zachariah Allen, manuscript autobiography, no date, in the Zachariah Allen papers, Rhode Island Historical Society (RIHS). Allen wrote that the president of an insurance company, to whom he applied for a rate reduction, told him, "I can not go about to see all the mills insured by me, and attend to my business at the office; and an average must be made. The good mills must pay for the poor." On rising rates as a motivation, Dane Yorke, *Able Men*, 233.

10. Z. Allen manuscript autobiography, Zachariah Allen papers, RIHS.

11. The Manufacturers Mutual Fire Insurance Company (incorporated, Chapter 34, Massachusetts of Acts of 1834) revived as a factory mutual company some twenty-five years later. In 1861, Mechanics Mutual Fire Insurance Company, chartered to insure houses, merged with the inactive Manufacturers Mutual to become Worcester Manufacturers Mutual and joined the AFM companies (Charles Nutt, *History of Worcester and Its People* [New York: Lewis Historical Publishing Company, 1919], 1027, and D. Hamilton Hurd, *History of Worcester County, Massachusetts* [Philadelphia: J. W. Lewis, 1889], 1553–54. Thanks to the Worcester Historical Museum Library for these references).

12. This was the Rhode Island Mutual Fire Insurance Company (Dane Yorke, *Able Men*, 26).

13. Hawthorne Daniel, *The Hartford of Hartford* (New York: Random House, 1960), 165.

14. Allen to James Read of Boston; letter reprinted in Dane Yorke, *Able Men*, 233–34.

15. Information on rates from Hawthorne Daniel, *The Hartford of Hartford,* 90; *The Factory Mutuals 1835–1935;* Dane Yorke, *Able Men*, 101; Philadelphia Board of Fire Underwriters, "Classes of Hazards and Rates of Premium . . . ," June 18, 1852. Atkinson stated that when MMFIC of Providence started, mill rates were 2 percent (*Proceedings of the Society of the Arts* at MIT, meeting 246, January 8, 1880, 22). Edward Atkinson, *The Prevention of Loss by Fire, Fifty Years' Record of Factory Mutual Insurance* (Boston: Damrell & Upham, 1900).

16. Spencer Kimball writes that some customers of mutual fire insurance companies (not factory mutuals) skipped town before paying their assessments (*Insurance and Public Policy* [Madison: University of Wisconsin Press, 1960], 43).

17. *The Factory Mutuals 1835–1935,* chart on p. 353.

18. Dane Yorke, *Able Men*, 137.

19. *The Chronicle* 81 (December 16, 1905): 1698–99.

20. On AFM's early standards, Frederick Down, "History of Factory Fire Insurance," in John Brown, ed., *Lamb's Textile Industries* (Boston: James H. Lamb, 1911), 248. On Francis's work, Edward V. French, *Factory Mutual Insurance . . . Compiled to Observe the 50th Anniversary of the Arkwright Mutual Fire Insurance Company* (Boston, privately printed, 1912), 8–9.

21. For example, in 1826, the Proprietors required all buildings over ten feet to have walls of brick or stone and a roof covered with slate or other noncombustible material ("Proposals by the Proprietors of the Locks and Canals on the Merrimack River, for the Sale of the Mill Power and Land at Lowell," Boston, 1826).

22. James B. Francis, "On the Means Adopted in the Factories at Lowell, Massachusetts, for Extinguishing Fire," *JFI* 49, 3rd series (April 1865): 268–75.

23. Charles Young, *Fires, Fire Engines, and Fire Brigades* (London: Lockwood, 1866), 41 and 44.

24. Richard Candee, "Early New England Mill Towns of the Piscataqua River Valley," in John Garner, ed., *The Company Town* (New York: Oxford University Press, 1992), 121.

25. James Montgomery, *A Practical Detail of the Cotton Manufacture of the United States* (Glasgow: John Niven, 1840). An article in an engineering reference book, published in London in 1836, described a floor made of "planks . . . about three inches thick, jointed and ploughed on the edges for the purpose of receiving slips of iron called tongues; this makes a tight and substantial floor." Such floors had been used in the Manchester area, according to the article, showing that some British owners followed the advice in the 1825 magazine article (Luke Hebert, *The Engineer's and Mechanic's Encyclopaedia*, vol. 1 [London: Thomas Kelly, 1836], 526).

26. Richard Candee dates the first appearance of plank floors to the late 1820s and believes that the idea came from an article in the London *Mechanics' Magazine* (4 [1825]: 8–9), which was reprinted in *American Mechanics Magazine* (Richard Candee, "The 1822 Allendale Mill and Slow-Burning Construction," *Journal of the Society for Industrial Archeology* 15 [1989]: 21–34). However, the 1825 article does not recommend the plank-style floor for fire resistance, but simply as a sturdy kind of floor. An Englishman who built a plank floor "about the year 1824" in his malt house recalled, "this floor, though somewhat novel in its construction, was simply made for strength, and without any idea of its being less liable to be destroyed by fire than ordinary floors" (reprinted in *Scientific American* supplement 4, no. 82 [July 28, 1877]: 1297). As to why owners built plank floors initially, if not for fire resistance, we have fragmentary evidence from Robert Israel, a mill overseer. In 1831, he described the floors in the new mills at Lowell as being "made without joists, by laying plank on the girders and a flooring of boards upon them—this will present an even surface underneath which will be painted, and is intended to do away with plastering, which is continually shaking down and injuring the machines." (Letter of Robert Israel to his uncle, Lewis Waln, August 9, 1831, Waln papers, Pennsylvania Historical Society. Thanks to Steven Lubar for his transcription and to Bruce Laverty of the Philadelphia Athenaeum for identifying Lewis Waln.)

27. Dane Yorke, *Able Men*, 235 and 83.

28. Edward V. French, *Factory Mutual Insurance*, 12–13.

29. Charles J. H. Woodbury, *The Fire Protection of Mills* (New York: John Wiley & Sons, 1882). Date of the first Report No. 5, "Mill Construction," which I have been unable to find, from Dane Yorke, *Able Men*, 184.

30. According to Woodbury, "twice as many fires are put out by pails as by any other means" (*AABN* 17 [February 28, 1885]: 97).

31. On Whiting, Dane Yorke, *Able Men*, 107–14, 145–46. On the origin of the laboratory, Dane Yorke, *Able Men*, 195. John Ordway, "Experiments on Lubricating Oils," *Proceedings, New England Cotton Manufacturers Association* (October 30, 1878). BMMFIC's laboratory grew into a major research enterprise, today's Factory Mutual Research Corporation.

32. Quote from C. J. H. Woodbury, *The Fire Protection of Mills*, 105.

33. BMMFIC, *Supplementary Report to January 1, 1880*. Atkinson noted that a few buildings in Boston—the post office and Mutual Life Insurance Company building—had standpipes.

34. J. R. Freeman, "Experiments Relating to Hydraulics of Fire Streams," ASCE *Transactions* 21 (November 1889): 307.

35. Woodbury's lecture, 1886–87, "Reminiscences of Benj. G. Buttolph Covering his Apprenticeship in the Inspection Dept. from July 31, 1888 to Nov. 30, 1891," JRF papers, MIT Archives. Woodbury's Cornell lecture was published as "The Evolution of the Modern Mill," *Scientific American* supplement no. 647 and no. 648 (1888): 10329-31 and 10346-47. On Freeman, JRF to Henry Evans, Vice President of Continental Insurance Company, May 5, 1894, JRF Papers, MIT Archives, box 29, Fire Insurance Company's Correspondence: 1894–1908.

36. C. J. H. Woodbury, *The Fire Protection of Mills*, 137.

37. Edward Atkinson, *The Fire-Engineer, the Architect, and the Underwriter: Their Relations to Each Other* (Boston, 1880), 17 and 18; C. J. H. Woodbury, "Methods of Reducing Fire Loss," 27.

38. *AABN* 5 (April 26, 1879): 129–30. As to "C"'s identity, my best guess is Francis W. Chandler, a Boston architect.

39. *AABN* 5 (May 10, 1979): 151–52. Atkinson was no more satisfied in this respect in 1888 (see *AABN* 24 [October 13, 1888]: 174–75). However, he felt no prejudice against the profession as such: his son William became an architect.

40. *AABN* 5 (May 24, 1879): 167.

41. *AABN* 5 (June 7, 1879): 182.

42. *AABN* 6 (August 16, 1879): 56.

43. Edward Atkinson, Report No. 5, "Slow-burning or Mill Construction," Insurance Engineering Experiment Station, September 1902. Circulars seem to have

been free until 1902, when Atkinson began to charge a subscription or small fee, in order to raise money for a testing laboratory and to endow a new course at MIT.

44. *AABN* 8 (November 27, 1880): 253.

45. Edward Atkinson, *AABN* 5 (June 6, 1979): 183. While not identified, the school was probably the Chauncy Hall School, built in the mid-1870s, about which Atkinson had been consulted (EA to Fire Marshal, City of Boston, March 2, 1894, EA letters, MHS, January 1, 1894–May 26, 1894).

46. Letter from W. G. Preston, *AABN* 24 (November 3, 1888): 211–12; *AABN* 18 (July 4, 1885): 6–7; and *42nd Annual Report of the Trustees of the Massachusetts School for the Feeble-Minded at South Boston, for the year ending Sept. 30, 1889.*

47. MIT, Cornell, and Woods Hole: Atkinson to W. H. Lincoln, November 18, 1886, EA letters, MHS (November 9, 1886–February 26, 1887); letter from Atkinson, *AABN* 24 (November 17, 1888): 236. Harvard Medical School, John R. Freeman, "Comparison of English and American Types of Factory Construction," *Journal of the Association of Engineering Societies* 10 (1890). Waltham hospital, EA to Edgar Moore, Rutland Board of Trade, February 13, 1894, EA letters, MHS, volume January 1, 1894–May 26, 1894. Atkinson's mailings to fire victims, *AABN* 24 (October 13, 1888): 175. The Yarmouth house, at 74 Main Street, was built for wood-pulp mill owner George W. Hammond ("A Mill-Built Dwelling," *AABN* 31 [January 31, 1891]: 74–75; Frank A. Beard et al., *Maine's Historic Places* [Camden, Maine: Down East Books, 1982], 345). On the Brown house, *AABN* 24 (November 17, 1888): 235.

48. Peter Wight, *AABN* 5 (June 7, 1879): 179. C. John Hexamer, "Fire Hazards in Textile Mills," *JFI* 119 (May 1885): 372–95. *The Architectural Era* 4 (1890): 234.

49. The Brown Building, Charles Woodbury, ASME *Transactions* 4 (November 1882–June 1883): 401. Atkinson on Boston architecture firms, Harold Williamson, *Edward Atkinson*, 125, from a letter written in 1886. "Lawrence Building, New York, N.Y.," *AABN* 41 (October 4, 1884): 162–63.

50. *IA* 2 (August 1883): 98.

51. Edward Atkinson, "The Prevention of Loss by Fire and the System of Factory Mutual Insurance, an Address . . . ," September 17, 1885, EA Papers, MHS. Leonard K. Eaton, *Gateway Cities and Other Essays* (Ames: Iowa State University Press, 1989), 11.

52. *AABN* 19 (April 24, 1886): 193 and 19 (May 22, 1886): 250–51. Letter of John E. Whitney, *AABN* 19 (May 22, 1886): 250–51.

53. Clifford Thomson, "The Waste by Fire," *The Forum* 2 (1886): 37.

54. Clifford Thomson, "The Waste by Fire," 36. C. J. H. Woodbury, "The Evolution of the Modern Mill," in *Scientific American* and also "Methods of Reducing Fire Loss," *Cassier's Magazine* 1 (November 1891–April 1892): 27–35 and 125–35. The second article was an update of one he read at an ASME meeting and pub-

lished in their *Transactions* 11 (1889–90): 271–310. Edward Atkinson, "Slow-Burning Construction," *The Century Magazine* 37 (1889): 566–79.

55. Edward Atkinson, "The Relation between the Architect and the Underwriter," *Proceedings of the Society of Arts* meeting 246, January 8, 1880, 22. Atkinson had been calculating this tax since 1880, at least. Other writers used the concept, but as far as I can tell, Atkinson was the first. See also Edward Atkinson, *Supplementary Report to January 1, 1880*, BMMFIC, 15.

56. *AABN* 26 (December 28, 1889): 297. Some people used the term *slow-burning* to describe buildings that did not conform to the New England standard. William J. Fryer Jr., long involved in New York City building regulation, explained a proposed revision to the law that would allow "increased height for non-fire-proof buildings if constructed on what is termed the slow-burning principles of filling up or cutting off the air spaces between the wooden floor beams" ("The New York Building Law," *Architectural Record* 1 [1891–92]: 81). This, of course, does not capture the full extent of the system. F. C. Moore, *Economical Fire-Resisting Construction*, rev. ed., May 1892 (first published 1890). Although he favored wood, the construction system he described was semifireproof, not slow-burning. In 1890, *AABN*'s editor compared Moore's publication with Atkinson's and Woodbury's work, apparently unaware of the differences between the systems (*AABN* 29 [September 20, 1890]: 173).

57. Quote from W. S. Eames, a St. Louis architect, in William Le Baron Jenney, "The Best Fireproof Construction for Buildings Occupied for Mercantile Purposes," *IA* 30 (October 1897): 24. Chicago pamphlet published in *IA* 20 (August 1892): 5. On lower rates for slow-burning construction, *AABN* 64 (May 13, 1899): 56.

58. Philadelphia: Pennsylvania laws of 1903, Act 236. Chicago, section of building ordinance quoted in William Le Baron Jenney, "The Best Fireproof Construction for Buildings Occupied for Mercantile Purposes," 23. Other cities, Thomas Nolan, editor-in-chief, Frank Kidder's *The Architects' and Builders' Pocket-Book*, 16th ed. (New York: John Wiley & Sons, 1916), 812.

59. Massachusetts laws of 1892, ch. 419, sections 41, 58, 63, and 64.

60. *Building Code Recommended by the National Board of Fire Underwriters*, 4th ed. (1915), section 76.

61. Frank E. Kidder, *The Architect's and Builder's Pocket-Book* (New York: John Wiley & Sons, 1885), chapter 24.

62. *AABN* 20 (November 27, 1886): 258–59.

63. "Special Report of the Fire Marshal of the City of Boston on the Kingston-Street Fire, Nov. 28, 1889," City Doc. No. 17, 1890; *AABN* 26 (December 14, 1889): 280. The fire started in a large warehouse at Kingston and Bedford Streets called the Brown-Durrell Building.

64. *AABN* 26 (December 7, 1889): 261.

65. H. C. Burdett letter to *AABN* 26 (December 7, 1889): 272.

66. For example, *AABN* 7 (January 31, 1880): 33; *AABN* 26 (December 28, 1889): 297.

67. Edward Atkinson, "Lessons From the Boston Fire," *AABN* 26 (December 21, 1889): 293–95.

68. For example, the architect John A. Fox concluded, "'Mill construction' has steadily grown in favor, but it is somewhat difficult of adaptation to complicated structures, and is chiefly useful in buildings attacked by fire from within" (*AABN* 26 [December 14, 1889]: 281–82).

69. *AABN* 40 (April 8, 1893): 30.

70. For example, a writer described the semifireproof Lumber Exchange building in Minneapolis as being "of the most approved slow-burning construction" (*IA* 18 [August 1891]: 7–10).

71. Edward Atkinson, "A Department of Insurance Engineering a Public Necessity," BMMFIC (March 1, 1902), 7.

72. *The Spectator* 49 (December 22, 1892): 296–97 (quote on 297), (December 29, 1892): 308, and 50 (January 5, 1893): 9. The articles originally appeared in the firefighter's magazine *Fire and Water*.

73. *AABN* 39 (March 18, 1893): 161.

74. *AABN* 40 (April 1, 1893): 14–15. That many understood mill construction simply as a way of building floors, see, for example, the chapter on "Slow-Burning Construction" in the series "Office-Help for Architects," *AABN* 44 (April 21, 1894): 27–28.

75. *Fireproof Magazine* 5 (October 1904): 6. On raising rates on "slow-burning" buildings that lacked sprinklers, W. L. B. Jenney, "The Best Fireproof Construction . . . ," 24.

76. *Fireproof Magazine* 2 (March 1903): 5–6.

77. *Fireproof Magazine* 2 (February 1903): 29.

78. Peter McKeon, *Fire Prevention* (New York: The Chief Publishing Company, 1912), preface.

79. Insurance Engineering Experiment Station, Report No. 7, "Fire-Resistant Roofs for Foundries and Machine Shops," 1903 and Report No. 10, "Tests of Columbian Fireproof Floors," 1903.

80. Edward Atkinson, "Concrete Construction," *Cement Age* 1 (April 1905): 463–67. E. A. Trego acknowledges Atkinson's contribution in Robert Lesley, ed., *Concrete Factories* (New York: Bruce & Banning, c. 1907), 101–2.

81. *Fireproof Magazine* 8 (1906): 11–15.

82. On beam size, e.g., Union Paper Cup Company, Trenton, N.J., described in Robert Lesley, ed., *Concrete Factories*, 83. On cost of concrete, George Maurice, "The Comparative Cost of Slow-Burning Mill and Reinforced Concrete Construction," *Cement Age* 1 (January 1905): 298–303. Other cost comparisons, A. P. Stradling in T. Nolan, ed., *The Architects' and Builders' Pocket-Book*, 16th ed. (1916), 758 and 772. On the end of Report No. 5, J. Crnkovich, "The New England Mutuals' Influence. . . ."

83. "History of Automatic Fire Extinguishing Systems," *The Factory Mutuals 1835–1935*, 229–48.

84. A rubber cap in the head served as a stopper. It was held in place by a rod that rested on a plug of solder at the bottom of the head. When the solder melted, the rod fell, releasing the rubber seal.

85. Charles Young, *Fires, Fire Engines, and Fire Brigades* 48, and Charles J. H. Woodbury, "Automatic Sprinklers," *Cassier's Magazine* 1 (November 1891–April 1892), 263. On the use of sprinklers in English factories, John R. Freeman, "Comparison of English and American Types of Factory Construction" (1890) and *AABN* 79 (January 10, 1903): 10. On English insurance companies, Charles J. H. Woodbury, "Methods of Reducing Fire Loss," 131.

86. Parmelee's patent no. 156,374, filed September 30, 1874. Charles Woodbury, in his history of the development of the automatic sprinkler, describes Parmelee's 1878 sprinkler head as if it was the 1874 head. But patent records indicate that Parmelee's first sprinkler (no. 154,076, filed June 24, 1874) had a head that used a spring. On Fall River, Edward V. French, *Factory Mutual Insurance*, 42. On the distrust of mill owners, Charles J. H. Woodbury, "Report on Automatic Sprinklers," BMMFIC *Special Report No. 19*, Boston, May 15, 1884.

87. 1878 patent, no. 205,672, filed December 27, 1878; "Highlights of Grinnell History," pamphlet in "Plumbing-Fire Protection" curator's files, Division of the History of Technology, National Museum of American History. On later developments, Everett Crosby et al., *Crosby-Fiske-Forster Hand-Book of Fire Protection*, 6th ed. (New York: D. Van Nostrand, 1919), 420–21.

88. Charles J. H. Woodbury, "Report on Automatic Sprinklers" and BMMFIC, *Special Report No. 10*, Boston, April 1, 1882, 30.

89. On Freeman's invention, *The Factory Mutuals 1835–1935*, 242. On the success of sprinklers, Charles J. H. Woodbury, "Automatic Sprinklers," 376.

90. On low rates as a disincentive, see comments following Edward Atkinson's paper, "The Relation between the Architect and the Underwriter," 30. On leakage policies, Harold Williamson, *Edward Atkinson*, 119–20.

91. "An Act, to provide for stability of construction and security against conflagration, panic, and other accident in theatres . . . ," *AABN* 6 (September 6, 1879): 77.

92. J.A.F., "American Dramatic Theatres," *AABN* 6 (September 6, 1879): 75. I

am guessing that J.A.F. is John A. Fox, a Boston architect who designed several theaters.

93. John R. Freeman, *On the Safeguarding of Life in Theaters* (privately printed, 1906), 34.

94. *AABN* 1 (December 23, 1876): 411.

95. Quote, *AABN* 19 (March 27, 1886): 145. New York's law: William J. Fryer Jr., *Law Relating to Buildings, in the City of New York* (New York: The Record and Guide, 1885), 49. Boston's law: Massachusetts laws of 1885, ch. 374, section 137. Sprinklers douse fires: Edward Atkinson letter, *AABN* 35 (January 23, 1892): 62. Sprinklers in theaters in 1905, J. R. Freeman, *On the Safeguarding of Life in Theaters*, 37.

96. On Cincinnati: *AABN* 33 (August 15, 1891): 99. New England Insurance Exchange: E. U. Crosby, *Hand-book of the Underwriters' Bureau of New England* (Boston: Standard, 1896). Joseph K. Freitag, *Fire Prevention and Fire Protection*, 2nd ed. (New York: John Wiley & Sons, 1921), 852–55.

97. Joseph K. Freitag, *Fire Prevention and Fire Protection*, 869.

98. Reported in *AABN* 16 (November 15, 1884): 229.

99. *The Factory Mutuals 1835–1935*, 97–98. E. Atkinson letter to Mr. Ballard, December 23, 1886, EA Papers, MHS, letter book November 9, 1886–February 26, 1887.

100. Edward Atkinson to Henry Deyer, March 29, 1894, EA Papers, MHS letter book, January 1, 1894–May 26, 1894. Industrial Risk Insurers, *The Hi-Spots* (February 1990): 3. IRI is the successor to FIA.

101. Harold Williamson, *Edward Atkinson*, 132–33. Edward Atkinson, "The Protection of City Warehouses from Loss by Fire" (February 1, 1883), EA Papers, MHS. Town mutual companies had a long history, and in the 1880s and 1890s, towns in the Midwest formed their own mutual fire companies. However, a conflagration could wipe out a town and its insurance system all at once, and this was not the kind of mutual Atkinson recommended. On town and church mutuals, see H. Roger Grant, *Insurance Reform: Consumer Action in the Progressive Era* (Ames: Iowa State University Press, 1979), chapter 4 and p. 96.

102. Edward Atkinson, "The Prevention of Loss by Fire ... an Address ...," 17. E. Crosby, *Hand-book of the Underwriters' Bureau of New England*, 1896.

103. National Fire Protection Association, Articles of Association, from 1911. W. H. Merrill, "Underwriters' Laboratories, Inc.," *The Spectator*, May 1, 1913.

104. Henry H. Hall, *Proceedings of the 38th Annual Meeting*, National Board of Fire Underwriters (May 1904), 18–19.

105. Letter from the Committee of Twenty, December 22, 1904, and minutes of

the Executive Committee, April 27, 1905, "Minutes of the Executive Committee, 1898–1906," ms, NBFU (now American Insurance Association).

106. *Proceedings of the 39th Annual Meeting*, National Board of Fire Underwriters (May 1905), 100.

107. F. W. Arnold and J. B. Branch, both officers in Providence, Rhode Island-based companies, *Proceedings of the 39th Annual Meeting*, National Board of Fire Underwriters, 1904, 108–11.

108. *Fireproof Magazine* 8 (1906): 90. This article also reported that the fire spread quickly because the building had open stair and elevator shafts, which would be inconsistent with mill construction.

109. "Detailed Studies of Fireproof Buildings in the Baltimore Conflagration," *Engineering News* 51 (February 25, 1904): 171.

Chapter 5. The Fireproof Skyscraper

Fireproof Magazine 8 (1906), p. 132.

1. Henry A. Goetz, in a paper read before the National Association of Fire Engineers, quoted in *IA* 16 (October 1890): 31; "Conflagrations Involving Property Losses of $500,000 and Over . . . ," *Proceedings of the 38th Annual Meeting* (NBFU, 1904), 38–39.

2. On passenger elevators, see Anne Millbrooke, "Technological Systems Compete at Otis," in William Aspray, ed., *Technological Competitiveness* (New York: IEEE, 1993), 243–69. On rents, William J. Fryer Jr., "A Review of the Development of Structural Iron," *A History of Real Estate, Building, and Architecture in New York City* (1898; New York: Arno Press, 1967), 464.

3. Owners of the Bennett Building, a fireproof office building in New York City that opened in 1873, promoted it as being "especially adapted to insurance, money brokers, merchants, lawyers, . . . businesses where security from fire to legal documents is a desideratum" (*New York Herald*, 1873, quoted in Gale Harris, "Bennett Building," N.Y. City Landmarks Preservation Commission, 3). Also in 1873, the American Institute of Architects rented rooms in a fireproof building to avoid "the possible loss by fire of our books and other property" (AIA, *Proceedings of the Seventh Annual Convention*, 12). On the Montauk, Peter Brooks letter of March 1881, quoted in Earle Shultz and Walter Simmons, *Offices in the Sky* (Indianapolis: Bobbs-Merrill, 1959), 24. Undoubtedly, few of the hotels that claimed to be "absolutely fireproof" were anything of the kind. Joseph K. Freitag wrote that owners advertised their ordinary buildings as fireproof "in an endeavor to increase their renting value" (*The Fireproofing of Steel Buildings* [New York: John Wiley & Sons, 1899], 24).

4. New York laws of 1860, ch. 470, sections 25 and 32. New York City's 1860 law was the first modern building code in the United States, in that it compiled and

superseded all previous rules related to buildings and also for the first time placed enforcement in the hands of men with experience in the building trades rather than fire-wardens (*Architects' and Mechanics' Journal* 2 [May 12, 1860]: 51; Joseph McGoldrick et al., *Building Regulation in New York City, A Study in Administrative Law and Procedure* [New York: The Commonwealth Fund, 1944], 45–46).

5. New York laws of 1867, ch. 939, section 32.

6. New York laws of 1871, ch. 625, section 28.

7. For example, "Tall Buildings in Chicago," *Iron Age* 48 (October 22, 1891): 691.

8. On fire fighting, John A. Fox paper, reprinted in *AABN* 26 (December 14, 1889): 282. Around 1880, the Committee on the Construction of Buildings of the National Board of Fire Underwriters gave a similar figure—seventy-five feet—as the maximum effective height for a fire stream (extracts of a report printed in *AABN* 8 [August 28, 1880]: 105). As of 1892, only the following cities called for standpipes on certain tall buildings: Chicago, Cincinnati, Denver, District of Columbia, Milwaukee, Providence, St. Louis, and San Francisco (Henry A. Phillips, compiler, "Comparative Municipal Building Laws—XXIV," *AABN* 39 (March 25, 1893) and "Comparative Municipal Building Laws—XXVI," *AABN* 40 [May 13, 1893]).

9. C. C. Knowles and P. H. Pitt, *The History of Building Regulation in London 1189–1972* (London: Architectural Press, 1972). Boston's preconflagration building law limited the heights of tenement houses on narrow streets to thirty feet. The postconflagration law of 1872 limited building heights overall to seventy-five feet, but this provision must have been set aside since buildings higher than this did go up (Massachusetts laws of 1871, ch. 280, section 35, and Massachusetts laws of 1872, ch. 371, section 20).

10. *IA* 1 (February 1883): 11.

11. Article IX of chapter 15 of the Municipal Code, section 1062½, Chicago; mentioned in *IA* 3 (April 1884): 33. (Thanks to Lyle Benedict, Municipal Reference Collection, for sending me the text.)

12. Massachusetts laws of 1885, ch. 374. New York laws of 1885, ch. 456, reprinted in William J. Fryer Jr., *Law Relating to Buildings, in the City of New York* (New York: The Record and Guide, 1885), 33. New York laws of 1887, ch. 566, section 492, reprinted in William J. Fryer Jr., *Laws Relating to Buildings, in the City of New York* (New York: The Record and Guide, 1887), 40. The cap was raised to eighty-five feet in 1892 and then in 1896, dropped back to seventy, only to be raised to seventy-five the following year.

13. C. W. Trowbridge, "Paper on Ironwork," *IA* 9 (June 1887): 77.

14. Charles J. H. Woodbury, "The Evolution of the Modern Mill," *Scientific American* supplement no. 648 (1888): 10347.

15. Paul Starrett, *Changing the Skyline: An Autobiography* (New York: Whittlesey House, 1938), 37.

16. Described in *IA* 5 (July 1885).

17. William J. Fryer Jr., *Law Relating to Buildings in the City of New York* (1885), 11. These dimensions were for the party walls of nonresidential buildings; residential buildings were allowed to have walls one brick less thick, and the front and rear walls of nonresidential buildings could be one brick less thick than the party walls.

18. On the Home Insurance building, Gerald Larson and Roula Geraniotis, "Toward a Better Understanding of the Evolution of the Iron Skeleton Frame in Chicago," *Journal of the Society of Architectural Historians* 46 (March 1987): 39–48. The interior court wall of the New York Produce Exchange (1881–84), designed by George B. Post, was an example of a skeleton wall; Post brought iron columns and girders out to the wall plane and filled between them with brick (*Engineering Record* 34 [July 11, 1896]: 103). Similarly, according to Peter Wight—who was the tile floor contractor for the building—the frame at the back side of the Phoenix Insurance Building in Chicago, completed in 1886, carried the walls (*IA* 19 [March 1892]: 22).

19. William J. Fryer Jr., "A Review of the Development of Structural Iron," 466.

20. William H. Birkmire, *Skeleton Construction in Buildings*, 3rd ed. (New York: John Wiley & Sons, 1900), 3.

21. William J. Fryer Jr., "A Review of the Development of Structural Iron," 464–65.

22. William J. Fryer Jr., "A Review of the Development of Structural Iron," 467, and William J. Fryer Jr., *Laws Relating to Building, in the City of New York* (1887). "Leading Building Material Firms," *A History of Real Estate . . . in New York*, 427.

23. William J. Fryer Jr., "Skeleton Construction, A New Method of Constructing High Buildings," *The Architectural Record* 1 (1891–92): 228–35 and also by Fryer, "A Review of the Development of Structural Iron," 471. J. C. Cady & Company designed the Lancashire Fire Insurance Company Building.

24. Ralph Peck, *History of Building Foundations in Chicago*, University of Illinois Engineering Experiment Station, Bulletin Series No. 373 (Urbana: University of Illinois Press, 1948).

25. Frank Randall, *History of the Development of Building Construction in Chicago* (Urbana: University of Illinois Press, 1949), 118, 120, 124. On Loring's suggestion, H. J. Burt, "Growth of Steel Frame Buildings; Origin and Some Problems of the Skyscrapers," *Engineering News-Record* 92 (April 17, 1924): 681. (Thanks to Cecil Elliott for this reference.) William Jenney designed the Manhattan building; Jenney & Mundie designed the Siegel, Cooper & Co. and Ludington.

26. Quote of Joseph K. Freitag in Thomas Nolan, ed., *The Architects' and Builders' Pocket-Book*, 16th ed. (New York: John Wiley & Sons, 1916), 234. Fryer considered the World a "notable example of cage construction," in *History of Real Estate . . . in New York*, 465–66. George Post, "Steel-Frame Building Construction," *Engineering Record* 32 (June 15, 1895): 44; also comments on "Wind Bracing in High Buildings," same number, 42; remarks, *AABN* 54 (November 28, 1896): 71–72. Capt. John S. Sewell of the Army Corps of Engineers was another engineer who advocated cage construction ("Lessons of the Baltimore Fire," *Proceedings of the 38th Annual Meeting*, NBFU, 1904, 135).

27. William Starrett, *Skyscrapers and the Men Who Build Them* (New York: Charles Scribner's Sons, 1928), 36.

28. Gustavs Henning, "Notes on Steel," ASME *Transactions* 4 (1882–83): 410. On the early steel industry, Peter Temin, *Iron and Steel in Nineteenth Century America* (Cambridge: MIT Press, 1964), 171–73.

29. William Sooy Smith, Chairman of the Committee on Tests of Iron and Steel of the ASCE, in "Message from the President of the U.S. transmitting the report of the board appointed to test iron, steel, and other metals . . . ," June 8, 1876 (Senate Exec. Doc. no. 71). On the need for information about steel, Robert H. Thurston, "Annual Address—the Mechanical Engineer," ASME *Transactions* 4 (1882–83): 76–77.

30. "Message from the President of the U.S. transmitting the report of the board appointed to test iron, steel, and other metals . . . ," June 8, 1876, 5.

31. Robert Thurston, *The Materials of Engineering, Part II. Iron and Steel* (New York: John Wiley & Sons, 1883), 372–75.

32. "The Cut in Beams," *Iron Age* 43 (January 10, 1889): 55.

33. Cooper Hewitt & Company to Samuel Reeves, Phoenix Iron Company, February 16, 1857, Phoenix Steel Corporation papers, Hagley Library, Manuscript Department.

34. "The Cut in Beams," *Iron Age*, 55.

35. *AABN* 24 (October 27, 1888): 189, with figures corrected in *AABN* 24 (November 17, 1888): 236. At first, *AABN* used the wrong number to convert kilograms to pounds. After correction, the editor's point still held: the specific duty was equal to the price of beams in France and Belgium.

36. *AABN* 26 (December 28, 1889): 297. Beams were more difficult to roll than rails, so a higher price was warranted.

37. Burton Hendrick, *The Life of Andrew Carnegie* (Garden City, N.Y.: Doubleday, Doran, 1932), 313.

38. *IA* 14 (January 1890): 94.

39. *IA* 16 (December 1890): 77.

40. *AABN* 35 (February 13, 1892): 97.

41. *AABN* 36 (June 4, 1892): 142.

42. Patent no. 35,582, June 17, 1862. The patent lists as its uses, "houses, piers, bridges & c." Alan Burnham, "The Rise and Fall of the Phoenix Column," *The Architectural Record* 125 (April 1959): 222–25; Christopher Baer, "Phoenix Steel Corporation," Hagley Library, Manuscripts and Archives department.

43. At an inquest on the mill collapse, James B. Francis, the eminent engineer from Lowell, Massachusetts, who later wrote a treatise on cast iron columns, testified that he had never before heard of an iron column breaking (*An Authentic History of the Lawrence Calamity* [Boston: J. J. Dyer, 1860], 96). On long columns, F. H. Kindl, ed., *Pocket Companion Containing Useful Information and Tables . . .* (Pittsburgh, 1896), 122.

44. Edward C. Shankland, "Steel Skeleton Construction in Chicago," *IA* 30 (January 1898): 56.

45. Frank Randall, *History of . . . Building Construction in Chicago*, 118–20, 122.

46. "Home Office Buildings of Distinction," *The Spectator* 129 (November 17, 1932): 13.

47. F. H. Kindl, *Pocket Companion Containing Useful Information and Tables*, preface. Of all the structural shapes rolled in 1889, roughly 276,000 gross tons, about 45 percent, were made of iron and the rest of steel. During the next decade, the tonnage of structural shapes more than tripled and in 1899, 97 percent of the total was made of steel (Peter Temin, *Iron and Steel in Nineteenth Century America*, table C9).

48. *A History of Real Estate . . . in New York*, 487 and 498.

49. Estimate by the engineer Alfred F. Evans, who condemned the use of cast iron columns (*New York Times*, March 13, 1904, 20).

50. *IA* 18 (August 1891): 7–11. Long and Kees designed the building.

51. *AABN* 38 (November 26, 1892): 132.

52. Quote of General William Sooy Smith and Isham Randolph, *AABN* 39 (January 14, 1893): 28; *IA* 20 (November 1892): 32; Joseph K. Freitag, *Fire Prevention and Fire Protection*, 2nd rev. ed. (New York: John Wiley & Sons, 1921), 135–37. Henry Ives Cobb designed the building.

53. Quotes, *IA* 13 (June 1889): 81. Peter Wight's pessimistic assessment in *AABN* 41 (August 19, 1893): 113.

54. Peter Wight, "Recent Fireproof Building in Chicago," Part IV, *IA* 19 (May 1892): 46.

55. "Tests of Fireproof Arches," *AABN* 31 (March 28, 1891): 195–201.

56. Peter Wight, "Recent Fireproof Building in Chicago," Part IV, 47.

57. Edward C. Shankland, "Steel Skeleton Construction in Chicago," 57. On Philadelphia, Edwin Betolett, "Fireproof Construction in Philadelphia," *Insurance Engineering* 2 (1901): 209.

58. Henry Maurer & Son, "Herculean Arch and Phoenix Wall Construction," (1908) trade catalogue.

59. Frank Kidder, *The Architect's and Builder's Pocket-Book*, 14th ed. (New York: John Wiley & Sons, 1906), 787–88. On tile floors in Europe, Fritz Von Emperger, "Hollow Tile Floors, Past & Present," ASCE *Transactions* 34 (1895): 521.

60. In the 1860s, the Missouri Concrete Stone Company produced "artificial stone" by the process patented by the English artificial stone manufacturer Frederick Ransome, for "bridges, gateways, mantels, vases, tiles, grave stones and every variety of ornamental work" (*Missouri Concrete Stone Co.* [St. Louis, 1867]. Thanks to Emily Miller, Missouri Historical Society, for sending a copy.). Reference to early use of concrete presses, *ARABJ* 1 (January 1868): 450. In 1868, four inventors patented presses for making concrete building blocks.

61. On Fox and Barrett, Lawrance Hurst, "The Age of Fireproof Flooring," in Robert Thorne, ed., *The Iron Revolution* (1990), 36. Phillips's floor consisted of a framework of small T-irons between rolled iron joists covered with concrete that included Portland cement (*Building News* 12 [December 29, 1865]: 925; *Building News* 22 [March 29, 1872]: 250). Dennett's was a plain concrete floor, usually arched and made of "Nottingham concrete," which consisted of gypsum and an aggregate (*Building News* 12 [April 7, 1865]: 243; G. M. Lawford, "Fireproof Floors," 45–46; Mike Williams with D. A. Farnie, *Cotton Mills in Greater Manchester* [Preston: Carnegie Publishing Company, 1992], 108). W. B. Wilkinson & Company's floor was an arch made of concrete that included Portland cement, broken brick, and ashes; the firm also made a flat style, for stairway landings and corridors, that used iron reinforcement (G. M. Lawford, "Fireproof Floors," 46). In 1862 Matthew Allen patented a floor with an iron framework similar to ones used in France. It consisted of a network of iron bars, with concrete made of Portland cement and cinders (G. M. Lawford, "Fireproof Floors," Society of Engineers *Transactions for 1889* [1890]: 52).

62. *AABN* 19 (March 6, 1886): 109. John Sewell, "Concrete and Concrete-Steel," *Cement Age* 1 (November 1904): 201. Jasper Draffin, "A Brief History of Lime, Cement, Concrete and Reinforced Concrete," in Howard Newlon, ed., *A Selection of Historic American Papers on Concrete 1876–1926* (Detroit: American Concrete Institute, 1976). U. Cummings, "Natural and Artificial Hydraulic Cements," *AABN* 22 (December 3, 1887): 269–71.

63. *ARABJ* 1 (January 1869): 450; E. L. Ransome and Alexis Saurbrey, "Reinforced Concrete Buildings" (1912), reprinted in Howard Newlon, ed., *A Selection of Historic American Papers on Concrete 1876–1926.*

64. *Fireproof* 2 (July 1903): 44.

65. "Concrete as a Building Material," *AABN* 2 (August 18, 1877): 266. Ward built the hazardous parts of his Port Chester Bolt Company factory out of "re-enforced béton" (W. E. Ward, "Béton in Combination with Iron as a Building Material," ASME *Transactions* 4 (November 1882–June 1883): 403).

66. Reprinted in Howard Newlon, ed., *A Selection of Historic American Papers on Concrete 1876–1926.* Hyatt's patent, no. 206,112 (July 16, 1878), covered the general idea of metal added to concrete to improve its tensile strength. Cyrille Simonnet, "The Origins of Reinforced Concrete," *Rassegna* 49 (March 1992): 11.

67. P. H. Jackson, "Iron and Concrete Construction," *AABN* 16 (October 4, 1884): 166; Walter Mueller, "Reinforced Concrete Construction," in Robert Lesley, ed., *Concrete Factories* (New York: Bruce & Banning, c. 1907), 64.

68. E. L. Ransome and Alexis Saurbrey, "Reinforced Concrete Buildings," in Howard Newlon, ed., *A Selection of Historic American Papers on Concrete 1876–1926.*

69. Furness patent no. 416,907, described in *The Architectural Era* 4 (June 1890): 130–32. On the American Philosophical Society, information from engineer Suzanne Pentz, letter, November 3, 1998.

70. On concrete manufacture, Robert Lesley, "The Manufacture of Cement," *Cement Age* 1 (October 1904): 145. On concrete output, Edwin Thacher, "Concrete and Concrete-Steel in the U.S.," *Cement Age* 1 (November 1904): 167. Concrete prices, Robert Lesley, *History of the Portland Cement Industry in the United States* (1924; New York: Arno Press, 1972).

71. Edward Shankland, "Steel Skeleton Construction in Chicago," 56–57. On concrete construction advantages and disadvantages, Thomas Nolan, ed., *The Architects' and Builders' Pocket-Book*, 16th edition (1916), 829.

72. The Bricklayers' Union opposed building code revisions that would have permitted concrete floors because they feared members would lose work (testimony of John J. Donnelly, business agent for Bricklayer's Union No. 7, in *Report of the Special Committee of the Assembly, Appointed to Investigate the Public Offices and Departments of the City of New York,* 1900 [cited hereafter as the Mazet Committee Report]).

73. The tests of cement and cement-mortar were made in Chicago, and the results published in 1896; described in Joseph K. Freitag, *The Fireproofing of Steel Buildings,* 96–97.

74. "Burnt Clay Fireproofing and Its Substitutes," *The Architectural Record* 8 (July–September 1898): 111–16.

75. "Failure and Efficiency in Fire-Proof Construction," *The Architectural Record* 7 (July 1897–July 1898): 393–98.

76. Joseph K. Freitag, *The Fireproofing of Steel Buildings,* 58–74; *The Engineering Record* 37 (September 18, 1897): 337–39.

77. William Tubby testimony, Mazet Committee Report, 1900.

78. Corydon Purdy, "Can Buildings be Made Fireproof?", ASCE *Transactions* 39 (June 1898): 141, quote on 159.

79. *AABN* 64 (April 15, 1899): 17.

80. Abraham Himmelwright, general manager of the Roebling Construction Company, testimony and Charles McCann testimony, Mazet Committee Report, 1900.

81. Francis C. Moore, *Fire Insurance and How to Build* (New York: Baker & Taylor, 1903), 20.

82. Quote from *Report of the Joint Committee of the Senate and Assembly of New York Appointed to Investigate . . . Insurance Companies . . .*, Assembly Doc. 30, February 1, 1911, 72–73. AABN's editors regularly criticized the stock fire insurance industry's practices. See, for example, *AABN* 22 (August 20, 1887): 81.

83. See the reports of the Committee on Statistics, 1860s through 1890s, in the *Proceedings* of the annual meetings of the National Board of Fire Underwriters.

84. Arthur C. Ducat, *The Practice of Fire Underwriting* (New York: T. Jones Jr., 1866), iii; Robert Riegel, "Problems of Fire Insurance Ratemaking," *Annals of the American Academy of Political and Social Science* 70 (March 1917): 199–219.

85. Henry A. Oakley, "Address," *Proceedings of the Eighth Annual Meeting*, NBFU (1874): 12–13.

86. State height limitation law, Massachusetts laws of 1891, ch. 355. Boston building laws, Massachusetts laws of 1892, ch. 419 and Massachusetts laws of 1893, ch. 464.

87. Chicago Real Estate Board, *Studies on Building Height Limitations in Large Cities* (Chicago: Chicago Real Estate Board Library, 1923), 14. For many years, Chicago's fire underwriters pressed for height limitations; in 1891, the local underwriters' association voted to refuse to insure buildings over 120 feet (*AABN* 35 [January 2, 1892]: 1).

88. The city's 1885 restriction on the height of apartment houses (eighty-foot maximum) had more to do with health concerns—preventing tall buildings from blocking the light of their neighbors in residential districts—than with fire safety. This limit remained in effect until 1897.

89. In 1913, many American cities had height limits on the books, however only in Boston, Chicago, and Washington, D.C., did many buildings approach the maximum. "The maximum height limit is no height limit at all so far as most buildings are concerned; it only prevents the erection of a few exceptionally high buildings" (Board of Estimate and Apportionment of the City of New York, *Report of the Heights of Buildings Commission*, December 23, 1913, 22).

90. The states were Georgia, Kansas, Ohio, Michigan, Missouri, Nebraska, New

Hampshire, and Texas (*Hayden's Annual Cyclopedia of Insurance in the United States 1899–1900* [Hartford: The Insurance Journal Company, 1900], 25–27).

91. William B. Clark, presidential address, *Proceedings of the Thirty-first Annual Meeting*, NBFU (May 1897): 21.

92. Quote, Francis C. Moore, *Fire Insurance and How to Build*, 697.

93. Estimates of the extent of damage vary from 1,500 to 2,500 buildings on seventy to eighty city blocks. Sources agree on property loss: $50 million. *New York Times*, February 8, 1904, 1, and February 9, 1904, 1; Roebling Construction Company, *The Baltimore Fire, The Iroquois Theatre Fire*, New York (c. 1904); National Fire Protection Association, *Conflagrations in America Since 1900* (Boston: NFPA, c. 1957).

94. Details from the Roebling Construction Company, *The Baltimore Fire, The Iroquois Theatre Fire*.

95. *New York Times*, February 9, 1904, 3.

96. Introduction to Charles L. Norton, *The Conflagration in Baltimore* (Boston: Insurance Engineering Experiment Station, 1904).

97. "Detailed Studies of Fireproof Buildings in the Baltimore Conflagration," *Engineering News* 51 (February 25, 1904): 169–73; Joseph K. Freitag, *Fireproof* 8 (1906): 174; *Municipal Journal and Engineer*, quoted in Mississippi Wire Glass Company, *A Reconnaissance of the Baltimore and Rochester Fire Districts . . .* (c. 1904), 5.

98. John R. Freeman, "An Engineer's Suggestions to Fire Underwriters," address at the annual banquet of the National Board of Fire Underwriters, May 12, 1904.

99. Roebling Construction Company, *The Baltimore Fire, The Iroquois Theatre Fire*, 22.

100. *Engineering News* 51 (February 18, 1904): 153.

101. Joseph K. Freitag, *The Fireproofing of Steel Buildings*, 26.

102. NFPA report quoted in Joseph K. Freitag, *Fire Prevention and Fire Protection*, 2nd rev. ed. (New York: John Wiley & Sons, 1921), 176. Charles L. Norton, *The Conflagration in Baltimore*, 10–11. "Detailed Studies of Fireproof Buildings in the Baltimore Conflagration," *Engineering News* 51 (February 25, 1904): 173.

103. From Richard L. Humphrey's report in G. K. Gilbert et al., *The San Francisco Earthquake and Fire of April 18, 1906*, Bulletin no. 324, Department of the Interior, USGS (Washington, D.C.: GPO, 1907), 49.

104. G. K. Gilbert et al., *The San Francisco Earthquake and Fire of April 18, 1906*, 64.

105. Joseph K. Freitag, *Fire Prevention and Fire Protection*, 180. John S. Sewell in

G. K. Gilbert et al., *The San Francisco Earthquake and Fire of April 18, 1906*, 68–69.

106. Joseph K. Freitag, *Fire Prevention and Fire Protection*, 181.

107. R. L. Humphrey in G. K. Gilbert et al., *The San Francisco Earthquake and Fire of April 18, 1906*, 53–54 and 58.

108. John S. Sewell in G. K. Gilbert et al., *The San Francisco Earthquake and Fire of April 18, 1906*, 120–21, and Frank Soulé, same book, 150. Joseph K. Freitag, *Fire Prevention and Fire Protection*, 181.

109. Henry Ericsson, *Sixty Years a Builder*, 282.

110. John S. Sewell, "Fire-Resistive Construction," chapter in Everett U. Crosby et al., *Hand-Book of Fire Protection* (New York: Van Nostrand, 1919), 128. On concrete in Baltimore, *Fireproof* 8 (1906): 39.

111. F. W. Fitzpatrick and Theodore L. Condron, *Fireproof Construction* (Chicago: American School of Correspondence, 1914), 47.

112. Board of Estimate and Apportionment of the City of New York, *Report of the Heights of Buildings Commission*, 276–77.

Chapter 6. Fire Exits

Quoted in *McClure's Magazine* 37 (September 1911), p. 482.

1. Frank Kidder, *The Architect's and Builder's Pocket-Book*, 14th ed. (New York: John Wiley & Sons, 1906), 726. Quote of Corydon T. Purdy in ASCE, *Transactions* 39 (June 1898): 145. Joseph K. Freitag, *Fire Prevention and Fire Protection*, 2nd rev. ed. (New York: John Wiley & Sons, 1921), 33.

2. Nicholas Wainwright, *The Philadelphia Story: The Philadelphia Contributionship* . . . (Philadelphia: The Philadelphia Contributionship, 1952), 42. Charlestown's law: Massachusetts laws of 1810, ch. 44. Brooklyn's law: New York laws of 1852, ch. 355, section 7. New York City: New York laws of 1862, ch. 356, sections 23 and 27. Boston: Massachusetts laws of 1871, ch. 280, section 30, required scuttles under a section that discussed roofing material; by 1872 (ch. 371, section 15), the scuttle rule followed one that called for fire escapes, which suggests it now was intended for egress.

3. C. C. Knowles and P. H. Pitt, *The History of Building Regulation in London 1189–1972* (London: Architectural Press, 1972), 63. R. E. H. Read, *British Fire Legislation on Means of Escape, 1774–1974*, Department of the Environment, Fire Research Station (Borehamwood, Hertfordshire, 1986).

4. *Architects' and Mechanics' Journal* 2 (May 12, 1860): 51. Joseph McGoldrick et al., *Building Regulation in New York City, A Study in Administrative Law and Procedure* (New York: The Commonwealth Fund, 1944), 46.

5. *Report of the Select Committee Appointed to Examine into the Condition of Tenant Houses in New-York and Brooklyn,* Assembly Doc. No. 205, Albany, 1857, 3; *New York Times,* February 4, 1860, 4.

6. Editorial, *New York Times,* February 4, 1860, 4. Account of fire, *New York Times,* February 3, 1860, 1.

7. *Architects' and Mechanics' Journal* 1 (March 17, 1860): 189.

8. New York laws of 1860, ch. 470. Whether the framers knew about the stair-towers in New England factories is uncertain, but they probably knew about the outside stair-tower at Harper & Brothers' famous building in New York City.

9. New York laws of 1862, ch. 356.

10. "Semi-annual Report of the Superintendent of Buildings," *Documents of the Board of Aldermen of the City of New York,* Part I, vol. 30, Doc. No. 7, February 26, 1863.

11. *AABN* 6 (October 4, 1879): 105.

12. New York laws of 1867, ch. 939.

13. New York laws of 1871, ch. 625, section 28.

14. Massachusetts laws of 1871, ch. 280, sections 19 and 37.

15. Massachusetts laws of 1872, ch. 371, sections 13 and 14.

16. *New York Times,* September 20, 1874, 1; September 21, 1; September 29, 1. Edward V. French, *Factory Mutual Insurance . . . Arkwright Mutual Fire Insurance Company* (Boston, 1912), 30.

17. *New York Times,* October 4, 1874, 5.

18. Ellen Jones vs. Granite Mills, Suffolk, December 24, 1878, *Cases Argued and Determined in the Supreme Judicial Court of Massachusetts,* November 1878–May 1879 (Boston: Little, Brown, 1880), 84–89; quote on 89.

19. *Sixth Annual Report of the Bureau of Statistics of Labor,* Massachusetts Public Doc. 31, March 1875, 145.

20. Massachusetts laws of 1876, ch. 214.

21. "General Specification for Fire-escapes . . . ," copy in *Tenth Annual Report of the District Police,* Massachusetts Public Document No. 32, 1888. Massachusetts laws of 1880, ch. 197.

22. "An Ordinance, Creating a board to regulate the construction and use of fire escapes," Philadelphia, December 15, 1876. (Thanks to Zandra Moberg, The Free Library of Philadelphia, for sending a copy.)

23. Laws of Pennsylvania, No. 132, "An Act To provide for the better security of life and limb, in cases of fire in hotels and other buildings," June 11, 1879.

24. William F. Willoughby, "Inspection of Factories and Workshops," *Mono-*

graphs on American Social Economics, U.S. Commission to the Paris Exposition of 1900.

25. Henry A. Phillips, comp., "Comparative Municipal Building Laws—XXIV," *AABN* 39 (March 25, 1893): 187.

26. Boston's 1885 building law, Massachusetts laws of 1885, ch. 374. Statewide law of 1888, Massachusetts laws of 1888, ch. 426. Boston's 1892 building law, Massachusetts laws of 1892, ch. 419, section 83.

27. Chicago's proposed theater ordinance authorized the building commissioner and fire marshal to set occupancy limits, which were to be part of a theater's license to operate, but it contained no practical method of monitoring compliance (*IA* 10 [January 1888]: 99).

28. William Gerhard, "Theatre Fire Statistics," *AABN* 46 (December 15, 1894): 115. The author of an early fire insurance textbook wrote that theaters, museums, and shows were uninsurable (Arthur Ducat, *The Practice of Fire Underwriting,* 4th ed. [New York: T. Jones Jr., 1866]). Francis C. Moore, of the Continental Insurance Company, advised agents to decline to insure theaters, explaining "they are unprofitable risks" (*Fires: Their Causes, Prevention and Extinction, . . . A Guide to Agents . . .*, July 1883, 243).

29. Mills's first big architectural commission, the Monumental Church in Richmond, was built in memory of the victims of this fire. Data on how theater fires started, William Gerhard, "Theatre Fire Statistics," 115–16. On fireproof wood, *The Mechanics' Magazine* 5 (1826): 60. William Strickland may have used cast iron columns to support the balcony in his Chestnut Street Theater in Philadelphia (1820–22) for their noncombustibility as much as for their slim profile.

30. *AABN* 1 (December 16, 1876): 402.

31. *AABN* 1 (December 16, 1876): 406–8. Thomas A. Jackson designed the building. *AABN* 1 (December 30, 1876): 423; *AABN* 6 (September 6, 1879): 75–77. William Fryer, *Law Relating to Buildings, in the City of New York* (New York: The Record and Guide, 1885), 40–50, section 500. Fryer, in a footnote, suggests that the architect Frank Kimball, a prominent theater designer, wrote the law. However, the theater section of the 1885 was virtually the same as the text of the earlier AIA bill. Kimball served on the original committee. Perhaps Kimball was responsible for such changes as were made in the final bill.

32. Massachusetts laws of 1885, ch. 374, sections 126 to 139.

33. Chicago correspondent, *AABN* 23 (March 24, 1888): 136. Boston's law: Massachusetts laws of 1892, ch. 419, section 96.

34. Hugh Bonner and Lawrence Veiller, *Tenement House Fire Escapes in New York and Brooklyn,* prepared for the Tenement House Commission of 1900 (New York, 1900).

35. Robert De Forest and Lawrence Veiller, *The Tenement House Problem*, vol. 1 (New York: Macmillan, 1903).

36. Frank Kidder recommended a rise of seven to seven-and-a-half inches (*The Architect's and Builder's Pocket-Book* [1906], 1476).

37. A survey of fire escapes in Pennsylvania factories found women especially could not get to the fire escapes ("Fire Prevention Study Given by the Alumnae . . . of Bryn Mawr College to the Pennsylvania Department of Labor and Industry," bound in *Second Annual Report of the Commissioner of Labor and Industry of the Commonwealth of Pennsylvania*, part II, 1915).

38. Everett U. Crosby et al., *Crosby-Fiske-Forster Hand-Book of Fire Protection* (New York: D. Van Nostrand, 1919), 558. Laws of the General Assembly of the Commonwealth of Pennsylvania, session of 1899, No. 123, section 38.

39. *Iron Age* 47 (April 16, 1891): 733. Single-family homes were excepted. A few years before, in October 1888, the architect George Frederick read a paper at the annual AIA convention in which he proposed the standard of a stairway for every 2,500 square feet of floor space (*IA* 12 [October 1888]: 39). There were no precedents for this idea in the building laws in Britain at this time.

40. Bureau of Buildings, Borough of Manhattan, *The Building Code of the City of New York, with amendments to April 12, 1906*, New York, 1906, sec. 75.

41. Robert De Forest and Lawrence Veiller, *The Tenement House Problem*, vol. 2, quote on 109 and 172.

42. Henry Ericsson, *Sixty Years a Builder* (1942; New York: Arno Press, 1972), 298.

43. Roebling Construction Company, *The Baltimore Fire, The Iroquois Theatre Fire* (c. 1904), quote on 54. John R. Freeman, *On the Safeguarding of Life in Theaters* (privately printed), 12.

44. Roebling Construction Company, *The Baltimore Fire, The Iroquois Theatre Fire*, 54.

45. John R. Freeman, "On the Safeguarding of Life in Theaters; Being a Study From the Standpoint of an Engineer," ASME *Transactions* 27 (1906): 71–170 and privately printed.

46. Roebling Construction Company, *The Baltimore Fire, The Iroquois Theatre Fire*, 60. Edwin O. Sachs, *The Fire at the Iroquois Theatre, Chicago* (London: The British Fire Prevention Committee, 1904).

47. "The New Ordinances for Chicago," section 65 for building classes IV and V, reprinted in Edwin O. Sachs, *The Fire at the Iroquois Theatre, Chicago*, 19–34.

48. *New York Times*, March 26, 1911, 1.

49. Report of H. F. J. Porter, industrial engineer, in *Preliminary Report of the Factory Investigating Commission*, State of New York, Senate Report No. 30, March 1, 1912, vol. 1, 160.

50. As found by the Joint Board of Sanitary Control's Fire Protection Investigating Committee, reported in the *New York Times,* March 27, 1911, 1:2.

51. *New York Times,* March 26, 1911, 1.

52. Alfred E. Smith, *Up to Now—An Autobiography* (New York: Viking, 1929), 87–93.

53. *Preliminary Report of the Factory Investigating Commission,* 823–26.

54. Peter McKeon, *Fire Prevention* (New York: The Chief Publishing Company, 1912), 9.

55. Articles of Association, NFPA, *Proceedings of the 15th Annual Meeting* (May 1911): 5.

56. NFPA, *Proceedings of the 15th Annual Meeting* (May 1911): 61.

57. NFPA, *Proceedings of the 18th Annual Meeting* (May 1914): 68–69.

58. "Report of Committee on Safety to Life," NFPA, *Proceedings of the 27th Annual Meeting* (1923).

59. Charles G. Smith, National Board of Fire Underwriters, *Proceedings of the Thirty-seventh Annual Meeting* (May 1903): 77–78.

60. Charles G. Smith, National Board of Fire Underwriters, *Proceedings of the Fortieth Annual Meeting* (May 1906): 75.

61. National Board of Fire Underwriters, *Building Code Recommended by the National Board of Fire Underwriters,* 4th ed., 1915.

62. Robert De Forest and Lawrence Veiller, *The Tenement House Problem,* vol. 2, xxi.

63. "Report of Committee on Fireproof Construction . . . ," NFPA, *Proceedings of the 17th Annual Meeting* (May 1913): quote on 133.

Infrastructure of Safety

"Fire Prevention Study Given by the Alumnae . . . of Bryn Mawr College to the Pennsylvania Department of Labor and Industry," bound in *Second Annual Report of the Commissioner of Labor and Industry of the Commonwealth of Pennsylvania* part II, 1915, p. 36.

1. Michael J. Karter Jr., "Fire Loss in the United States in 1997," *NFPA Journal* 92 (September–October 1998): 74.

2. Merritt Roe Smith, *Harpers Ferry Armory and the New Technology* (Ithaca: Cornell University Press, 1977).

3. The idea that regulations stifle innovation is not borne out by research. Francis T. Ventre, in *Social Control of Technological Innovation,* examined the effect of regulation on technological innovation in home building. He concluded that it did not have a negative impact, even though the slow pace of code revision frus-

trates individuals seeking changes (Ph.D. diss, Massachusetts Institute of Technology, 1973).

4. *AABN* 38 (November 5, 1892): 18.

5. Horace Cubitt, "A Comparison of English and American Building Laws," *AABN* 89 (May 19, 1906): 180.

6. *NFPA Journal* 92 (September–October 1998): 74–82; U.S. Bureau of the Census, *Statistical Abstract of the U.S.: 1996* (Washington, D.C., October 1996), table 361. Fatality figures exclude firefighters.

Glossary

Arch   A structure that crosses an opening. Although arches are usually curved, flat structures that support loads mainly by compression rather than by bending are also called arches.

Beam   A horizontal structure that crosses an opening that supports loads mainly by bending. A large beam is usually called a girder. A smaller beam that supports a floor is called a joist.

Béton   French word for concrete.

Centering   Temporary wooden framework that makes a base for constructing brick or stone arches, or concrete arches and slabs.

Concrete   A mixture of a bonding material, such as lime or cement, along with water, sand, and a large-sized aggregate, such as stones and coal ashes.

Deafening   A layer of material, usually concrete or mortar, placed between joists or between layers of a floor to muffle sound. When applied thickly, this layer also served as a fire barrier.

End-construction   Hollow block floor arches in which the cavity of the block is perpendicular to the beam.

Fire clay   A kind of clay mixed with sand and other infusible materials that has great heat resistance.

Fire prevention   Eliminating and reducing the causes of fire, such as keeping chimneys clean.

Fire protection   As applied to buildings, designing and equipping a building so that a fire on the inside can be controlled and one outside cannot break in. Constructing a firewall to divide the floor area of a large building is a fire protection measure.

Flange   Part projecting from a beam or column. On an I-beam, for example, the horizontal top and bottom of the beam are the flanges.

Girder   A large horizontal structure that crosses an opening.

Groin   Usually defined as the curved edge at the juncture of two intersecting vaults.

Joist   A horizontal structure that is part of a floor frame.

Pier   A vertical support made of masonry, which can be freestanding or part of a wall. Wall material concentrated between window openings, carrying the floor loads, was called a pier.

Portland cement   Hydraulic cement made according to a recipe.

Reinforced concrete   Concrete structures in which metal (iron or steel) ropes, bars, or wires have been placed in order to increase the structure's tensile strength.

Side-construction   Hollow block floor arches in which the cavity of the block is parallel with the beam.

Skewback   The structural units at the end of an arch, in which the side of the block that faces the arch is slanted so as to start the springing of the arch.

Soffit   The underside of a structural member, such as the bottom surface of the lower flange of an I-beam.

Standpipe   In buildings, a vertical water pipe used to supply fire-hose outlets.

Terra cotta   Clay.

Tile   Clay products, with the exception of bricks. The words *terra cotta* and *tile* are synonymous.

Truss   An assembly of tension and compression members (e.g., beams, posts, diagonal rods, and struts) that perform the same function as a deep beam.

Vault   An arched structure that forms a floor or roof and is made of masonry.

Voussoir   Wedge-shaped blocks, with sides slanted toward a central key block, that form an arch.

Web   The vertical part of a beam or the walls in a hollow tile block.

Bibliographical
Essay

Histories of Fireproof Construction
and the Impact of Fires

The best histories of American fireproof construction were written by the men who made it, notably the engineer Joseph K. Freitag; foundry manager, engineer, architect, and building official William J. Fryer Jr.; and the architect and manufacturer Peter B. Wight. Freitag's books, in addition to presenting a view of best practice at the turn of the century, also provide an historical overview of fireproof construction practice. I used Freitag's *Architectural Engineering*, 2nd ed. (New York: John Wiley & Sons, 1904); *Fire Prevention and Fire Protection, As Applied to Building Construction*, 2nd ed. (New York: John Wiley & Sons, 1921); and *The Fireproofing of Steel Buildings* (New York: John Wiley & Sons, 1899). Wight's valuable chronicles of fireproof construction and hollow tile include "On the Relation of Architecture to Underwriting," *AABN* 5 (1879), serialized; "Recent Fireproof Building in Chicago," *IA* 5 (April 1885) and *IA* 19 (1892), serialized; "Origin and History of Hollow Tile Fire-Proof Floor Construction," *The Brickbuilder* 6 (1897), serialized. Fryer's overviews include "A Review of the Development of Structural Iron," in *A History of Real Estate, Building and Architecture in New York City* (1898; New York: Arno Press, 1967) and "Skeleton Construction, A New Method of Constructing High Buildings," *The Architectural Record* 1 (1891–92): 228–35.

Useful textbooks on fireproof construction include F. W. Fitzpatrick and Theodore L. Condron, *Fireproof Construction* (Chicago: American School of Correspondence, 1914) and Frank Kidder, *The Architect's and Builder's Pocket-Book* (New York: John Wiley & Sons), first published in 1885. Kidder's books, regularly revised and reissued, allow the reader to see when new materials were introduced.

Several British architects and engineers wrote overviews of fireproof building materials used in Great Britain which provide useful comparisons with American practice. These surveys include Arthur Cates, "Concrete and Fire-Resisting Construction," RIBA *Sessional Papers* (1877–78): 296–309; G. M. Lawford, "Fireproof Floors," Society of Engineers (London) *Transactions for 1889* (1890): 43–70; and John J. Webster, "Fire-proof Construction," Institution of Civil Engineers *Min-*

utes of Proceedings 12 (1890–91): 249–88. Two valuable modern studies of fire-proof materials used in Britain are S. B. Hamilton, *A Short History of the Structural Protection of Buildings Particularly in England*, National Building Studies, Special Report No. 27, Department of Scientific and Industrial Research (London: HMSO, 1958), and Lawrance Hurst, "The Age of Fireproof Flooring," in Robert Thorne, ed., *The Iron Revolution*, Essays to Accompany an Exhibition at the RIBA Gallery (London, 1990). While these authors catalogue the variety of methods introduced over time, they do not say how extensively any of them were used. Of course, this is a very hard question to answer.

There is little secondary material on the history of fireproof construction in the United States. Carl Condit discusses fireproof buildings briefly in *American Building: Materials and Techniques*, 2nd ed. (Chicago: University of Chicago Press, 1982), and again in a more recent book, written with Sarah Landau, *The Rise of the New York Skyscraper 1865–1913* (New Haven: Yale University Press, 1996). He does not treat fireproof construction as a system. Donald Friedman, in his book *Historical Building Construction* (New York: W. W. Norton, 1995), presents an overview of high-tech building materials and methods used in New York City, which covers fireproof construction. Aimed at designers and engineers who work on old buildings, it focuses more on what transpired rather than why.

The question of how building regulation influences the development of building technology has received little study. One treatment of the topic, although not historical, is Francis T. Ventre, *Social Control of Technological Innovation: The Regulation of Building Construction* (Ph.D. diss., MIT, 1973). His focus was on homebuilding, rather than the more technologically dynamic commercial/industrial construction. Marian Browley, in *The British Building Industry* (Cambridge: Cambridge University Press, 1966), asked whether London's height restrictions stifled innovation in commercial construction, and she concluded that they did not.

I found two useful studies of the impact of fire on urban growth: John Weaver and Peter Lottinville, "The Conflagration and the City: Disaster and Progress in British North America During the Nineteenth Century," *Histoire sociale-Social History* 13 (November 1980): 417–49; L. E. Frost and E. L. Jones, "The Fire Gap and the Greater Durability of Nineteenth Century Cities," *Planning Perspectives* 4 (September 1989): 333–47. Because I found her conclusions unpersuasive, I did not use Christine Rosen's book on redevelopment in three cities following major fires, *The Limits of Power* (Cambridge: Cambridge University Press, 1986).

Statistics on fires and discussions of fire protection standards are contained in the fire protection handbooks, proceedings of annual meetings, and journals published by the National Fire Protection Association. A useful overview of America's contemporary fire situation is John Hall Jr. and Arthur Cote, "America's Fire Problem and Fire Protection," in Arthur Cote, ed., *Fire Protection Handbook*, 18th ed. (Quincy, Mass.: NFPA, 1997). The NFPA's *Conflagrations in America Since 1900* (Boston: NFPA, c. 1957) contains a list of great fires.

Periodicals

Architectural and engineering periodicals contain valuable information on fire-proof construction practice. The ones I relied on the most are *American Architect and Building News, Architects' and Mechanics' Journal, The Architectural Record, Architectural Review and American Builders' Journal, The Brickbuilder, Engineering News, Engineering Record, Fireproof Magazine,* and *Inland Architect.* I found several trade magazines to be useful, including *The Iron Age* for information on the iron industry and use of iron; *The Cement Age* on concrete construction; and *The Spectator* on the fire insurance industry. For information on foreign influences on American design, I mainly used the British architectural magazine *Building News.* I also found articles on practice, and often transcriptions of interesting debates and comments, in the annual publications of professional/trade organizations: American Institute of Architects, American Society of Civil Engineers, National Board of Fire Underwriters, and National Fire Protection Association.

Chapter 1
The Solid Masonry Fireproof Building,
1790–1840

To write chapter 1, I relied on biographies of early American architects, the archives of Robert Mills (available on microfilm, *The Papers of Roberts Mills, 1781–1855,* Wilmington, Del.: Scholarly Resources, 1990), contemporary city guidebooks, historic structures reports, and personal inspections of extant buildings. The biographies included Jeffrey A. Cohen and Charles E. Brownell, *The Architectural Drawings of Benjamin Henry Latrobe,* vol. 2, part 1 (New Haven: Yale University Press, 1994); Helen M. P. Gallagher, *Robert Mills, Architect of the Washington Monument, 1781–1855* (1935; New York: AMS Press, 1966); Agnes Gilchrist, *William Strickland, Architect and Engineer* (Philadelphia: University of Pennsylvania Press, 1950); Constance Greiff, *John Notman, Architect, 1810–1865* (Philadelphia: Athenaeum of Philadelphia, 1979); Talbot Hamlin, *Benjamin Henry Latrobe* (New York: Oxford University Press, 1955); Roger Newton, *Town & Davis, Architects* (New York: Columbia University Press, 1942); William Wheildon, *Memoir of Solomon Willard* (The Monument Association, 1865); and Edward Zimmer, *The Architectural Career of Alexander Parris* (Ph.D. diss., Boston University, 1984). Information on the first vaulted fireproof building was available in Negley Teeters, *The Cradle of the Penitentiary; The Walnut Street Jail at Philadelphia, 1773–1835* (The Pennsylvania Prison Society, 1955).

Chapter 2
The Iron and Brick Fireproof Building,
1840–1860

The history of the I-beam in chapter 2 is based on material from a number of manuscript collections: Records of the Public Building Service (Record Group 121) at the National Archives; records of the Architect of the Capitol's office;

Cooper Hewitt & Company collection in the Manuscript Division of the Library of Congress; and Phoenix Steel Corporation records at the Hagley Museum and Library in Wilmington, Delaware. Record Group 121 contains correspondence of the supervising architect and records of all buildings constructed under the Treasury Department's control. I used principally entry 6, "Letters Sent, Chiefly by the Supervising Architect, September 1852–August 1862"; entry 26, "Letters Received, 1843–1910," containing incoming correspondence, contracts, reports, and financial records for individual buildings, by year; and entry 90, "Construction Contracts 1854–1860." The Treasury Department's Office of Construction collected drawings of its 1850s buildings in volumes entitled "Plans of Public Buildings in the Course of Construction Under Direction of the Secretary of the Treasury," 1855–56. Sets of these drawings can be found in the National Archives, Library of Congress, and many other libraries as well as on microfilm.

At the Architect of the Capitol's Office, I looked at records pertaining to the construction of the Patent Office, Washington post office, and U.S. Capitol in the 1850s.

The Cooper Hewitt & Company collection at the Library of Congress contains record books for the 1850s to 1870s, some of which were from the Trenton office of the Trenton Iron Company and some from the New York office of Cooper Hewitt & Company. I looked at daybooks, order books, and shipping records—whatever I could find, in order to have information covering every year from 1852 to 1873. Unfortunately, no such material is available from Phoenix Iron Company; only fragmentary information survives in the Phoenix Steel Corporation records at the Hagley Library from the early years of beam rolling. Still, this collection contains some useful nineteenth-century material on Phoenix columns.

The important primary sources on iron construction include William Fairbairn, *On the Application of Cast and Wrought Iron to Building Purposes* (New York: John Wiley, 1854); Eaton Hodgkinson, *Experimental Researches on the Strength and Other Properties of Cast Iron* (London: John Weale, 1846); and Thomas Tredgold, *A Practical Essay on the Strength of Cast Iron* (London, 1822). My study updates the classic scholarly articles on the development of the wrought iron I-beam. These include two by Robert Jewett, "Structural Antecedents of the I-Beam, 1800–1850," *Technology and Culture* 8 (July 1967): 346–62, and "Solving the Puzzle of the First American Rail-Beam," *Technology and Culture* 10 (July 1969): 371–91; and two by Charles Peterson, "Inventing the I-Beam: Richard Turner, Cooper & Hewitt and Others," in H. Ward Jandl, ed., *The Technology of Historic American Buildings* (Washington, D.C.: Foundation for Preservation Technology, 1983), and "Inventing the I-Beam, Part II: William Borrow at Trenton and John Griffen of Phoenixville," APT *Bulletin* (1994): 17–25. Esmond Shaw's study, *Peter Cooper and the Wrought Iron Beam* (New York: Cooper Union, School of Art and Architecture, 1960), also contains many details on the beginnings of beam rolling.

Studies of cast iron construction that I found helpful include Turpin Bannister's articles, "The First Iron Framed Buildings," *Architectural Review* 107 (1950): 231–46; and in *Journal of the Society of Architectural Historians* "Bogardus Revisited, Part I: The Iron Fronts," 15 (December 1956): 12–22, and "Bogardus Revisited, Part II: The Iron Towers," 16 (March 1957): 11–19; James Dilts and Catharine Black, eds., *Baltimore's Cast-Iron Buildings and Architectural Ironwork* (Centerville, Md.: Tidewater, 1991); Ferdinand C. Latrobe, *Iron Men and Their Dogs* (Baltimore: Ivan Drechsler, 1941); and John Gloag and Derek Bridgewater, *A History of Cast Iron in Architecture* (London: George Allen and Unwin, 1948). Margot Gayle's introduction in *Badger's Illustrated Catalogue of Cast Iron Architecture* (New York: Dover, 1981) provides a useful introduction to cast iron in buildings.

Still the best economic history of nineteenth-century iron and steel is Peter Temin's *Iron and Steel in Nineteenth Century America: An Economic Inquiry* (Cambridge: MIT Press, 1964). In a chapter on cities in his book *A Nation of Steel* (Baltimore: Johns Hopkins University Press, 1995), Thomas Misa presents a useful overview of the development of building technology with respect to the market for structural steel. Jeanne McHugh's biography of Alexander Holley, *Alexander Holley and the Makers of Steel* (Baltimore: Johns Hopkins University Press, 1980), provides a good overview of the origin of the steel industry in the United States. Likewise, Alan Nevin's biography, *Abram Hewitt* (New York: Harper & Brothers, 1935), helped me see the development of the industry from the standpoint of this important ironmaster.

Regarding fireproof factories in Britain, about which a great deal has been written, good overviews of the topic can be found in Keith Falconer, "Fireproof Mills—The Widening Perspectives," *Industrial Archaeology Review* 16 (Autumn 1993): 11–26; Ron Fitzgerald, "The Development of the Cast Iron Frame in Textile Mills to 1850," *Industrial Archaeology Review* 10 (Spring 1988): 127–45; Colum Giles and Ian Goodall, *Yorkshire Textile Mills, 1770–1930* (London: HMSO, 1992); Mike Williams with D. A. Farnie, *Cotton Mills in Greater Manchester* (Preston: Carnegie, 1992). These update Jennifer Tann's still informative book, *The Development of the Factory* (London: Cornmarket, 1970).

Chapter 3
Response to the Great Fires

To write chapter 3, I made heavy use of the R. G. Dun & Company Collection (an early credit-reporting agency) at the Manuscripts Department, Baker Library, Harvard University Graduate School of Business Administration. The huge ledger books in this collection contain notes from the company's agents on the debts and financial prospects of firms doing business in New York. For data on patents, I mainly used the *Annual Report of the Commissioner of Patents*, a.k.a. *Official Gazette* of the U.S. Patent Office, which are available in government deposit libraries. These annual reports contain brief descriptions and illustrations, if applicable, of all patents issued during the year. Patents predating 1874 are indexed cumulatively by subject (*Subject-Matter Index of Patents for Inventions. . .*

1790–1873 [1874; New York: Arno, 1976]). For 1874 on, there are annual indexes.

The main sources of information on the history of hollow tile are Peter Wight's articles and Joseph K. Freitag's *Fireproofing of Steel Buildings*, cited above. Helpful introductions to the terra cotta and hollow tile industries can be found in Sharon Darling, *Chicago Ceramics and Glass, an Illustrated History from 1871 to 1933* (Chicago: Chicago Historical Society, 1979) and her *Decorative and Architectural Arts in Chicago, 1871–1933* (Chicago: University of Chicago Press, 1982).

To learn about architects' interest in fireproof construction in the important decade of the 1870s, I read the papers on the topic published by the American Institute of Architects in its *Proceedings of Annual Conventions*, and used archival material at the American Institute of Architects headquarters in Washington, that is, minutes of executive meetings and scrapbooks. The AIA *Proceedings* carried one on the earliest surveys of fireproof construction in America, Peter Wight's "Remarks on Fire-Proof Construction," *Proceedings of the Second Annual Convention* (1868), reprinted in *Architectural Review and American Builders' Journal* 2 (August 1869): 99–108. The AIA commissioned an important early guide intended to help architects use iron beams: Robert G. Hatfield, "Fireproof Floors for Banks, Insurance Companies, Office Buildings and Dwellings," AIA *Proceedings of the Second Annual Convention* (1868).

Chapter 4
Mill Fire Protection Methods
Enter the Mainstream

The main works I used to write chapter 4 were publications for and about the factory mutual fire insurance companies. Three excellent histories of the factory mutuals are Edward V. French, *Factory Mutual Insurance . . . Compiled to Observe the 50th Anniversary of the Arkwright Mutual Fire Insurance Company* (Boston, privately printed, 1912); Manufacturers Mutual Fire Insurance Company, *The Factory Mutuals 1835–1935* (Providence: Manufacturers Mutual Fire Insurance Company, 1935); and Dane Yorke, *Able Men of Boston . . . the . . . Story of . . . the Boston Manufacturers Mutual Fire Insurance Company* (Boston: BMMFIC, 1950). Edward Atkinson, the central character in this chapter, wrote voluminously, and a collection of his papers can be found at the Massachusetts Historical Society. Harold Williamson's biography, *Edward Atkinson, The Biography of an American Liberal 1827–1905* (Boston: Old Corner Bookstore, 1934), contains a bibliography of Atkinson's published writings. Of these, Atkinson's *The Prevention of Loss by Fire, Fifty Years' Record of Factory Mutual Insurance* (Boston: Damrell & Upham, 1900) and "Slow-Burning Construction," *The Century Magazine* 37 (February 1889): 566–79 provide an overview of factory mutual insurance and the development of slow-burning construction from his perspective. At the archives of the Massachusetts Institute of Technology, I used the papers of another important AFM official and civil engineer, John R. Freeman. Freeman's "Comparison of English and American Types of Factory Construction," *Journal of the Associa-*

tion of Engineering Societies 10 (1890) presents his views of the AFM prevention approach.

Important publications about mill fire protection history and practice by Charles J. H. Woodbury, another engineer who helped make it, include *The Fire Protection of Mills* (New York: John Wiley & Sons, 1882); "The Evolution of the Modern Mill," *Scientific American* supplement no. 647 and no. 648 (1888): 10329-31 and 10346-47; "Methods of Reducing Fire Loss," *Cassier's Magazine* 1 (November 1891–April 1892): 27-35 and 125-35; and "Automatic Sprinklers," *Cassier's Magazine* 1 (November 1891–April 1892): 261–67 and 369–77.

Betsy Bahr treated the evolution of mill architecture in her Ph.D. dissertation, *New England Mill Engineering: Rationalization and Reform in Textile Mill Design, 1790–1920* (University of Delaware, 1987). My main concern was how the AFM companies influenced the evolution of fire-resisting construction overall, not the development of building form specifically, which was Bahr's focus. However, I inevitably cover much of the same ground she did, especially her chapter 4. Another history of AFM and its influence on mill design is John J. Crnkovich, "The New England Mutuals' Influence on Industrial Architecture" (January 1978, unpublished; copy at the National Museum of American History).

Chapter 5. Triumph

No one has yet written a history of the fire insurance industry in the United States, but a history of the industry's national trade association comes very close: Harry C. Brearley, *Fifty Years of a Civilizing Force, An Historical and a Critical Study of the Work of the National Board of Fire Underwriters* (New York: Frederick A. Stokes, 1916). An article that people often cite concerning the history of the industry, although it does not contain much information, is F. C. Oviatt, "Historical Study of Fire Insurance in the United States," in Kailin Tuan, ed., *Modern Insurance Theory and Education*, vol. 1 (Orange, N.J.: Varsity, 1972). For information on fire insurance business practices in the nineteenth century, I relied on the National Board of Fire Underwriters, *Proceedings of Annual Meetings*. The NBFU is now called the American Insurance Association. Histories of fire insurance companies also proved helpful, particularly Hawthorne Daniel, *The Hartford of Hartford* (New York: Random House, 1960), and Henry Gall, *100 Years of Fire Insurance* (Hartford: Aetna Insurance Company, 1919). The underwriting genius, Francis C. Moore, president of the Continental Insurance Company in New York from 1889 to 1902, wrote several important books on fire insurance and construction, notably *Fire Insurance and How to Build* (New York: Baker & Taylor, 1903) and *Fires: Their Causes, Prevention and Extinction, . . . A Guide to Agents . . .* (New York, July 1883). In the early nineteenth century, several states conducted investigations of corruption in the fire insurance industry. New York State's investigation, *Report of the Joint Committee of the Senate and Assembly of New York Appointed to Investigate . . . Insurance Companies*, Assembly Doc. 30 (1911), contained much valuable information about the industry.

To trace the history of building construction in the dynamic 1880s and 1890s, I used two essential reference books, *History of Real Estate, Building and Architecture in New York City* (1898; New York: Arno Press, 1967) for New York City, and Frank Randall, *History of the Development of Building Construction in Chicago* (Urbana: University of Illinois Press, 1949). The former contains a list of building permits granted by year, and the latter contains descriptions and sources on all important buildings in downtown Chicago for the nineteenth century.

To see nineteenth-century building laws, one usually has to go to them. Sometimes private publishers put out copies of the laws, and these occasionally turn up outside the home city. William Fryer's annotated versions of New York City's building laws for 1885, 1892, and 1894, published by the Record and Guide, are available on microfilm in the American Architectural Books collection. In the nineteenth century, all building codes in Massachusetts and for New York City (before 1897) were chapters of state law and thus are available in the state libraries. For building codes in Portland, Oregon, I used copies in the city library and city archives; I obtained copies of relevant parts of Chicago's and Philadelphia's codes from the city libraries. I found copies of Seattle's and San Francisco's codes in the city libraries. A compilation of provisions in the building laws of America's large cities, prepared by Henry A. Phillips, was published in *American Architect and Building News* in 1893 as, "Comparative Municipal Building Laws," beginning in volume 33 (August 1, 1891) and running to volume 40 (May 13, 1893), serialized. Equally important were model building codes prepared by the National Board of Fire Underwriters, first published in 1905 and updated periodically, through the first half of the twentieth century. I used several different editions of these.

Fortunately, there are three helpful studies of the history of New York City's building laws: John P. Comer, *New York City Building Control, 1800–1941* (New York: Columbia University Press, 1942); William J. Fryer Jr., "The New York Building Law," *The Architectural Record* 1 (1891–92): 70–82; and Joseph McGoldrick et al., *Building Regulation in New York City, A Study in Administrative Law and Procedure* (New York: The Commonwealth Fund, 1944). At the turn of the century, New York City's Tenement House Commission published a report that contained excerpts of the city's nineteenth-century building laws: Robert De Forest and Lawrence Veiller, *The Tenement House Problem* (New York: Macmillan, 1903). For information on British building regulations, to put American rules in perspective, a good overview is C. C. Knowles and P. H. Pitt, *The History of Building Regulation in London 1189–1972* (London: Architectural Press, 1972).

On the building height limit movement, I used the Board of Estimate and Apportionment of the City of New York, *Report of the Heights of Buildings Commission*, New York, December 23, 1913, and Chicago Real Estate Board, *Studies on Building Height Limitations in Large Cities, with Special Reference to Conditions in Chicago* (Chicago: Chicago Real Estate Board Library, 1923). The views of an opponent of the limitation movement can be found in Earle Shultz and Walter Simmons, *Offices in the Sky* (Indianapolis: Bobbs-Merrill, 1959).

The best texts on the development of skeleton construction were William Birkmire, *Skeleton Construction in Buildings,* 3rd ed. (New York: John Wiley & Sons, 1900); Freitag's *Architectural Engineering* and *Fireproofing of Steel Buildings;* and Fryer's articles, "Skeleton Construction, A New Method of Constructing High Buildings" and "A Review of the Development of Structural Iron." There is a wealth of contemporary literature on the history and practice of skeleton construction. Good overviews can be found in Louis De Coppet Berg, "Iron Construction in New York City, Past and Future," *The Architectural Record* 1 (1891–92): 448–68; H. J. Burt, "Growth of Steel Frame Buildings: Origins and Some Problems of the Skyscrapers," *Engineering News-Record* 92 (April 17, 1924): 680–84; George B. Post, "Steel-Frame Building Construction," synopsis of a paper in *Engineering Record* 32 (June 15, 1895): 44; Corydon T. Purdy, "Can Buildings be Made Fireproof?," ASCE *Transactions* 39 (June 1898): 121–49; and Edward C. Shankland,"Steel Skeleton Construction in Chicago," *IA* 30 (January 1898): 56–57.

Reminiscences of builders proved extremely informative. I read Henry Ericsson, *Sixty Years a Builder* (1942; New York: Arno Press, 1972); Paul Starrett, *Changing the Skyline: An Autobiography* (New York: Whittlesey House, 1938); and Paul's better-known brother and sometime business partner, William Starrett, *Skyscrapers and the Men Who Build Them* (New York: Charles Scribner's Sons, 1928). An interesting study of the history of foundations was Ralph B. Peck, *History of Building Foundations in Chicago,* University of Illinois Engineering Experiment Stations Bulletin Series no. 373 (Urbana: University of Illinois Press, 1948). Secondary works on the history of structural engineering that I found useful include Henry J. Cowan, *Science and Building* (New York: John Wiley & Sons, 1978).

For information on fireproofing products, I used trade catalogues. I found the largest collections of the ones I wanted in the Warshaw Collection, Archives Center, Smithsonian Institution; library of the National Museum of American History; Hagley Museum and Library; and the Athenaeum of Philadelphia. The handbooks of the rolling mills, for example, F. H. Kindl, ed., *Pocket Companion Containing Useful Information and Tables . . .* (Pittsburgh, 1896), provided valuable information on structural design practice. For concrete, Robert Lesley's *Concrete Factories* (New York: Bruce & Banning, c. 1907) and his *History of the Portland Cement Industry in the United States* (New York: Arno Press, 1972) give much information about the history of the cement industry and concrete buildings. Among the important early accounts of concrete construction is William Ward, "Béton in Combination with Iron as a Building Material," ASME *Transactions* 4 (November 1882–June 1883): 388–403. Howard Newlon, ed., *A Selection of Historic American Papers on Concrete 1876–1926* (Detroit: American Concrete Institute, 1976) contains other important early papers on concrete. The transcript of the New York City investigation on corruption in the building department provides insight into contracting operations at the time, along with the politics of building laws: *Report of the Special Committee of the Assembly, Appointed to In-*

vestigate the Public Offices and Departments of the City of New York, January 15, 1900 (Mazet Committee Report).

Of the books and articles about the effects of the great conflagrations on buildings, a number stand out. For Baltimore: Charles L. Norton, *The Conflagration in Baltimore*, Report No. 13 (Boston: Insurance Engineering Experiment Station, 1904); Roebling Construction Company, *The Baltimore Fire, The Iroquois Theatre Fire*, New York, c. 1904; John S. Sewell, "Lessons of the Baltimore Fire," *Proceedings of the 38th Annual Meeting* National Board of Fire Underwriters (1904); and Mississippi Wire Glass Company, *A Reconnaissance of the Baltimore and Rochester Fire Districts . . .* (c. 1904). For San Francisco: G. K. Gilbert et al., *The San Francisco Earthquake and Fire of April 18, 1906*, Bulletin no. 324, Department of Interior, USGS (Washington, D.C.: GPO, 1907).

Chapter 6. Calamity

For chapter 6, I used early building codes and tenement house studies and laws to trace the history of emergency egress regulation. To compare American with British practice, I used R. E. H. Read, *British Fire Legislation on Means of Escape, 1774–1974*, Department of the Environment, Fire Research Station (Borehamwood, Hertfordshire, 1986), which contains excerpts of relevant laws. An early statement of the problem of egress in tenement houses can be found in *Report of the Select Committee Appointed to Examine into the Condition of Tenant Houses in New-York and Brooklyn*, Assembly Doc. no. 205, Albany, 1857. That adequate emergency egress remained a problem for tenants was documented by the Tenement House Commission of 1900 in Hugh Bonner and Lawrence Veiller, *Tenement House Fire Escapes in New York and Brooklyn* (New York, 1900).

For information on the Iroquois fire, I used John R. Freeman, *On the Safeguarding of Life in Theaters; Being a Study From the Standpoint of an Engineer*, reprinted privately from the ASME *Transactions* 27 (1906) and Roebling Construction Company, *The Baltimore Fire, The Iroquois Fire*, New York, c. 1904. Information on the Triangle fire, reports, and testimony taken in order to recommend worker protection legislation are contained in the *Preliminary Report of the Factory Investigating Commission*, State of New York, Senate Report No. 30, March 1, 1912. People who know about this fire have either read, or read sources that relied on, Leon Stein's account in *The Triangle Fire* (Philadelphia: J. B. Lippincott, 1962), in which he concludes that the deaths were the result of a crime: a locked door. This conclusion obscures the larger issue, which was that exits from the shop were so defective that some workers would have died in any event, even if both ninth-floor doors—and workers charged one was locked—had been unlocked. For an excellent contemporary overview of the problem of fires in high buildings, see Arthur McFarlane, "Fire and the Skyscraper," *McClure's Magazine* 37 (September 1911): 466–82.

Index

Hollow pots, 14, 85, 234n. 59
Hollow tile column fireproofing, 91,
    95–96, *96*
Hollow tile floors: chosen over con-
    crete blocks, 90, 247n. 68; compared
    with concrete, 168–69; cost of, 91;
    early use of, in U.S., 87–88, *89,*
    90–91, 248n. 74; end-construction,
    92, 93, 160–62, *161,* 249n. 84; Euro-
    pean use of, 85–86, 163, 245nn. 51
    & 52; French styles of, 87–88, *89;*
    origins of, in U.S., 78, 86, 243n. 20,
    245n. 53; porous tile in, 97; side-
    construction, 92, *161;* simplified
    styles of, 92–93, 162, *163;* superseded
    brick arches, 100; tests of, 91, 92–93,
    160–62
Hollow tile partitions, 88, 91
Hollow tile roof blocks, 91, 93, *94*
Home Insurance Building, 144–45,
    153
Houses of Parliament, London, 42
Hudson River Railroad Company
    depot, 74, 240n. 2
Humphrey, Richard, 183
Hunt, Richard M., 83, 90
Hupfels Brewery, 74
Hutton, Nathaniel H., 83
Hyatt, Thaddeus, 166

Illinois prison at Joliet, 74, 240n. 4
Illinois State Capitol, 90
Illinois Terra-Cotta Lumber Company,
    162, 163
I. M. Singer & Company building, 52
Insurance Engineering Experiment
    Station, 127, 181
Iron, properties of cast and wrought,
    57–58
Iron and brick arch fireproof con-
    struction, 37–38, *39,* 72; architec-
    tural consequences of, 71; British ar-
    chitects use of, 42–43, 79–80; with

cast iron frames, in U.S., 47, 48–52,
    53, 54–55; collapses of, 42–43; cost
    of, 41, 55, 72, 91; design of frame of,
    39–40, 44, 72; hazard of, 79–80; not
    used in American textile mills, 104;
    origins in Britain, 38–39, *40;* super-
    seded by hollow tile system, 100;
    tested in fires, 80, 81–82; weight of
    arches in, 39, 228n. 3; wrought iron
    in, 58–59, 62, *64,* 66–71
Iron grid frame filled with stone, 41,
    42, 53, 75, 228n. 5, 241n. 8
Iron output, U.S., 45, 46
Iron window sash, 22, 223n. 25
Iroquois Theatre fire, 203–5

Jackson, P. H., 166
Jenney, William Le Baron, 153–54
John J. Schillinger company, 101
John, James, 99
Johnson, George H., 52, 86–88, 92,
    93, 97, 99, 246n. 55

Kearsley Home building, 70
Kellogg Bridge Works, 102
Kendall building, 87–88, *89*
Kirby, Charles, 53
Kreischer, Balthasar, 86–87, 88, 89–
    90, 246n. 56

Ladders and platforms, 111, 194
Laing, David, 14
Lancashire Fire Insurance Company
    building, 148
Langley, Batty, 13
Latrobe, Benjamin Henry, 16–20, 23,
    25, 26, 28, 31, 35
Lee, Thomas A., 160–61
Lenox Library, 92
Lewis, T. Hayter, 79
Lienau, Detlef, 83
Lord & Taylor store, 83, 244n. 42
Loring, Sanford E., 93, 149, 249n. 89

Library of Congress Cataloging-in-Publication Data
Wermiel, Sara E.
  The fireproof building : technology and public safety in the
nineteenth-century American city / Sara E. Wermiel.
      p.   cm. — (Studies in industry and society ; 19)
  Includes bibliographical references and index.
  ISBN 0-8018-6311-2 (alk. paper)
    1. Building, Fireproof—United States—History—19th century.
2. Fires—United States—History—19th century. 3. Cities and
towns—United States—History—19th century. I. Title.
II. Series.
TH1065 .W47 2000
628.9′22′097309034—dc21   99-052228